The Palgrave Macmillan Animal Ethics Series

Series editors
Andrew Linzey
Oxford Centre for Animal Ethics
Oxford, UK

Priscilla Cohn
Penn State Abington
Villanova, PA, USA

Associate editor
Clair Linzey
Oxford Centre for Animal Ethics
Oxford, UK

Aims of Series

In recent years, there has been a growing interest in the ethics of our treatment of animals. Philosophers have led the way, and now a range of other scholars have followed from historians to social scientists. From being a marginal issue, animals have become an emerging issue in ethics and in multidisciplinary inquiry. This series will explore the challenges that Animal Ethics poses, both conceptually and practically, to traditional understandings of human–animal relations. Specifically, the Series will:

- provide a range of key introductory and advanced texts that map out ethical positions on animals;
- publish pioneering work written by new, as well as accomplished, scholars;
- produce texts from a variety of disciplines that are multidisciplinary in character or have multidisciplinary relevance.

More information about this series at
http://www.springer.com/series/14421

A.W.H. Bates

Anti-Vivisection and the Profession of Medicine in Britain

A Social History

A.W.H. Bates
Department of Cellular Pathology
University College London
London, UK

The Palgrave Macmillan Animal Ethics Series
ISBN 978-1-137-55696-7   ISBN 978-1-137-55697-4   (eBook)
DOI 10.1057/978-1-137-55697-4

Library of Congress Control Number: 2017939314

To Lucky,
A spark divine

The original version of the book was revised: The new section "Series Editors' preface" was included.

Series Editors' Preface

This is a new book series for a new field of inquiry: Animal Ethics.

In recent years, there has been a growing interest in the ethics of our treatment of animals. Philosophers have led the way, and now a range of other scholars have followed from historians to social scientists. From being a marginal issue, animals have become an emerging issue in ethics and in multidisciplinary inquiry.

In addition, a rethink of the status of animals has been fuelled by a range of scientific investigations which have revealed the complexity of animal sentiency, cognition and awareness. The ethical implications of this new knowledge have yet to be properly evaluated, but it is becoming clear that the old view that animals are mere things, tools, machines or commodities cannot be sustained ethically.

But it is not only philosophy and science that are putting animals on the agenda. Increasingly, in Europe and the United States, animals are becoming a political issue as political parties vie for the "green" and "animal" vote. In turn, political scientists are beginning to look again at the history of political thought in relation to animals, and historians are beginning to revisit the political history of animal protection.

As animals grow as an issue of importance, so there have been more collaborative academic ventures leading to conference volumes, special journal issues, indeed new academic animal journals as well. Moreover, we have witnessed the growth of academic courses, as well as university posts, in Animal Ethics, Animal Welfare, Animal Rights, Animal Law, Animals and Philosophy, Human–Animal Studies, Critical Animal Studies, Animals and Society, Animals in Literature, Animals and Religion—tangible signs that a new academic discipline is emerging.

"Animal Ethics" is the new term for the academic exploration of the moral status of the non-human—an exploration that explicitly involves a focus on what we owe animals morally, and which also helps us to understand the influences—social, legal, cultural, religious and political—that legitimate animal abuse. This series explores the challenges that Animal Ethics poses, both conceptually and practically, to traditional understandings of human–animal relations.

The series is needed for three reasons: (i) to provide the texts that will service the new university courses on animals; (ii) to support the increasing number of students studying and academics researching in animal related fields, and (iii) because there is currently no book series that is a focus for multidisciplinary research in the field.

Specifically, the series will

- provide a range of key introductory and advanced texts that map out ethical positions on animals;
- publish pioneering work written by new, as well as accomplished, scholars, and
- produce texts from a variety of disciplines that are multidisciplinary in character or have multidisciplinary relevance.

The new Palgrave Macmillan Series on Animal Ethics is the result of a unique partnership between Palgrave Macmillan and the Ferrater Mora Oxford Centre for Animal Ethics. The series is an integral part of the mission of the Centre to put animals on the intellectual agenda by facilitating academic research and publication. The series is also a natural complement to one of the Centre's other major projects, the

Journal of Animal Ethics. The Centre is an independent "think tank" for the advancement of progressive thought about animals, and is the first Centre of its kind in the world. It aims to demonstrate rigorous intellectual enquiry and the highest standards of scholarship. It strives to be a world-class centre of academic excellence in its field.

We invite academics to visit the Centre's website www.oxfordanimal-ethics.com and to contact us with new book proposals for the series.

Andrew Linzey
Priscilla N. Cohn
General Editors

Acknowledgements

This book was conceived during a sabbatical funded by the Wellcome Trust (grant number 104505/Z/14/Z). I am very grateful to them for their support and to Professor Joe Cain and the staff of the department of Science and Technology Studies at University College London for hosting me. In 2015, UCL kindly reappointed me as an honorary member of staff. As both they and the Welcome Trust are active supporters of experiments on animals, I would like to express my particular appreciation for their indulgence toward a dissenting view.

It was a pleasure to visit the London Metropolitan Archives, the National Archives at Kew, and the RSPCA Archive at Southgate. My thanks also to Ms Annie Lindsay of University College London Hospital Archives, and Mr Dan Mitchell of UCL Special Collections for their kind assistance.

The Reverend Professor Andrew Linzey has been an inspiration to me, and many other, students of animal ethics, and was kind enough to nominate me as a Fellow of the Oxford Centre for Animal Ethics, whose meetings have provided a wonderful forum for the exchange of ideas, and from which I have learned a great deal.

This project was made possible through the generous encouragement of the managing director of Convit House Pathology Ltd, and my wife, An (they are one and the same person), who kindly overlooked the fact that my duties both as a director, and husband, were not performed even to my usual low standard.

In recent years, the National Health Service has mandated my attendance at an ever-increasing number of superfluous meetings, on the way to and from, and sometimes during, which, much of this book was written.

Two chapters are reworked versions of articles written for the *Journal of Animal Ethics*: Chap. 1 is based on 'Vivisection, virtue ethics, and the law in nineteenth-century Britain', *J. Anim. Ethics*, 4;2 (2014), 30–44, and Chap. 4 on 'Boycotted Hospital: The National Anti-Vivisection Hospital, London, 1903–1935', *J. Anim. Ethics*, 6;2, 177–187.

Contents

About the Author

A.W.H. Bates is a Coroner's Pathologist, Honorary Senior Lecturer at University College London, and Fellow of the Oxford Centre for Animal Ethics. His other books include *Emblematic Monsters* (2005) and *The Anatomy of Robert Knox* (2010).

Abbreviations

AAMR	Association for the Advancement of Medical Research
ADAVS	Animal Defence and Anti-Vivisection Society
AFS	Animals Friend Society
BBC	British Broadcasting Corporation
BMA	British Medical Association
BMJ	British Medical Journal
BUAV	British Union for the Abolition of Vivisection
COS	Charities Organization Society
FRS	Fellow of the Royal Society
ICRF	Imperial Cancer Research Fund
LDAVS	London and District Anti-Vivisection Society
LMA	London Metropolitan Archives
LRCP	Licentiate of the Royal College of Physicians
MP	Member of Parliament
MRC	Medical Research Council
MRCS	Member of the Royal College of Surgeons
NCDL	National Canine Defence League
OGA	Order of the Golden Age
RDS	Research Defence Society
(R)SPCA	(Royal) Society for the Prevention of Cruelty to Animals
UCL	University College London

UFAW   Universities Federation for Animal Welfare
ULAWS   University of London Animal Welfare Society
VSS   Victoria Street Society
Well   Wellcome Library, London

Abstract

This book provides an introduction to the social history of the anti-vivisection movement in Britain from its nineteenth-century beginnings until the 1960s. Its focus is on the ethical principles that inspired the movement and the sociopolitical background that explains its rise and fall. Opposition to vivisection began when medical practitioners complained it was contrary to the compassionate ethos of their profession. Christian anti-cruelty organizations took up the cause out of concern that callousness among the professional classes would have a demoralizing effect on the rest of society. As the nineteenth century drew to a close, the influence of transcendentalism, Eastern religions and the spiritual revival led new age social reformers to champion a more holistic approach to science, and dismiss reliance on vivisection as a materialistic oversimplification. In response, scientists claimed it was necessary to remain objective and unemotional in order to perform the experiments necessary for medical progress.

1

Introduction

In Britain, the great majority of vivisection—I use the term not just in its literal sense of cutting up living animals but in the broader one of experimenting on them to the extent that they suffer and perhaps die—has been performed in the name of medicine.[1] It is of historical interest, therefore, that, in the mid-nineteenth century, when vivisection was introduced into British laboratories and medical schools from the Continent, much of the opposition to it came from doctors. Their motives in resisting it were complex, but can be boiled down to a conviction that vivisection was not something that a doctor ought to do.

Inspired by their example, this account of the history of the British anti-vivisection movement from its beginnings until the 1960s will focus on the character, or virtue, of the experimenter. This is not being done for the sake of finding a novel perspective on a topic that has been fairly well studied, but because I believe the virtue-centred approach was one of the nineteenth century's most significant contributions to the debate, and that it might profitably be revisited by present day animal ethicists. It may be no coincidence that the massive increase in animal experimentation in the twentieth century coincided with the rise of utilitarianism in medical ethics. As medical scientists assumed greater ethical autonomy, their emphasis was on results, eschewing

© The Author(s) 2017   1
A.W.H. Bates, *Anti-Vivisection and the Profession of Medicine in Britain,*
The Palgrave Macmillan Animal Ethics Series, DOI 10.1057/978-1-137-55697-4_1

conventional virtues on the one hand, and sentiment and feeling on the other, with dire consequences for laboratory animals.[2]

In writing a book for a series on animal ethics, I am labouring under a disadvantage in not being an ethicist. Most doctors are in the same position, and in the past 20 years or so a cadre of professional medical ethicists has arisen to guide our thinking. This has not been unproblematic, since few are trained in medicine and almost none practice it. While clinicians have benefited from the input of academic ethicists, some of their contributions have proved controversial, and occasionally shocking, such as the recent paper in a leading British journal of ethics that seemed, certainly to the readers who contacted the journal to complain, to be advocating infanticide. Its perspective was, needless to say, utilitarian.[3] The resulting outcry apparently surprised some members of the academic community, especially since, as the journal's editor observed, arguments in favour of infanticide had previously been put forward by 'the most eminent philosophers and bioethicists in the world…'.[4] Of course, it is not to be supposed that these luminaries would be anything other than horrified if their theorising actually led to children being killed, but the furore does reveal the gulf between ethical theory and practice, and, since the latter is influenced by the former, serves as a reminder that how ethics is taught matters. Bioethics as currently taught in British medical schools is unlikely to stress the importance of the physician's humane character[5]: as anyone who works in a teaching hospital will know, medical students and junior doctors are trained to seek the greatest benefit for the largest number; and to their utilitarian hammer, everything looks like a nail.

The physician's ethical grounding, or lack thereof, is important to the history of vivisection because, since the nineteenth century, by far its commonest defence has been that it saves human lives. It was on this basis that, in 1941, an English court made the landmark ruling that anti-vivisection was no longer to be considered a charitable cause because, as experiments on animals had been shown to be of benefit to humankind, anti-vivisectionists would be harming their fellow humans by banning them. This utilitarian judgement assumed that medics were justified in performing such experiments as were necessary to alleviate human suffering. Apart from a few conscientious objectors, most people working in British biomedicine today subscribe to this view, and will say, if asked, that they dislike experiments on animals, and hope alternatives will be found in time, but are currently obliged to accept them as the only way to develop life-saving treatments. Of course, since the publication of Peter Singer's influential *Animal Liberation* (1975), the interests of animals are less likely to be omitted from the utilitarian calculus, but few if any experimenters value animal lives on a level with human ones. Singer's own commitment to their interests did not extend to calling for the abolition of vivisection, and the balanced position he expressed in 1980 is very close to the view of the medical research community today:

> The knowledge gained from some experiments on animals does save lives and reduce suffering. Hence, the benefits of animal experimentation exceed the benefits of eating animals and the former stands a better chance of being justifiable than the latter; but this applies only when an experiment on an animal fulfils strict conditions relating to the significance of the knowledge to be gained, the unavailability of alternative techniques not involving animals, and the care taken to avoid pain. Under these conditions the death of an animal in an experiment can be defended.[6]

There is very little in this book on the subject of animal rights, since while the concept has a long history, going back at least as far as Thomas Tryon (1634–1703),[7] and while parallels between the rights of man and the rights of animals helped put animal welfare on the political agenda at the end of the eighteenth century, rights-based arguments were seldom used by the anti-vivisection movement before the 1970s. It was only after the publication of the Oxford philosophers' collection of essays, *Animals, Men and Morals* in 1971, followed by *Animal Liberation* (in which Singer himself does not take a rights-based position), that the anti-cruelty movement regrouped under the banner of 'animal rights', and began to appeal primarily 'to reason rather than to emotion or sentiment', a significant departure from what had gone before.[8] Rights-based arguments have certainly caught the public imagination, to the extent that non-specialist publications typically refer to anti-vivisectionists as 'animal rights activists' as though the two are synonymous, with the result that the principles on which the movement was founded are prone to be overlooked.

As Andrew Linzey has written, an idea that has generated as much scholarly activity as animal rights cannot be unimportant, but merely respecting any rights that animals are deemed to have falls far short of the ideal of treating them humanely, an ideal that was at the heart of the original anti-vivisection movement.[9] This history will explore the ethical arguments that sustained that movement from its beginning and throughout its heyday: namely, that it is socially irresponsible to permit cruelty, that Christianity, and other faiths, require animals to be treated as more than means to an end, and that a balanced, holistic approach to medicine must draw on emotional and spiritual insights as well as on the results of experiment.

It is possible that some will turn to this book for an answer to the question of whether animal experimentation has in fact resulted in significant medical advances. In this, as in other aspects of the debate, the propaganda has been extreme: supporters of vivisection have laid claim to almost every major medical advance of the twentieth century, while their opponents argue that nothing useful has been learned. The truth lies in between, though it is difficult to say precisely where, as just because a particular advance *was* made through vivisection does not mean that it could *only* have been thus made. Since vivisection has been normative, or even compulsory, in laboratories for over a century, it has necessarily played a part in most of the important discoveries made during that time, but whether they would have been made without it is as difficult to answer as any question in hypothetical history, and I shall not attempt to do so here. It is notable that historically, while the claim that vivisection has resulted in key advances in medicine has been a critical one for its supporters, for many anti-vivisectionists it was irrelevant, since the issue for them was whether vivisection benefited society as a whole, the answer to which depended on a sophisticated socio-political value judgement of what makes for a good society.

So prominent was the culture of the laboratory in the late-nineteenth century that trials on 'suffering human beings' were dismissed by the *British Medical Journal* (BMJ) as inferior to experiments on animals.[10] Like human dissection half a century earlier, vivisection was controlled by legislation, restricted to professionals, confined to licensed premises, and performed out of the public gaze, and though its practitioners were disliked and sometimes feared by the public, they were also admired for their fortitude and commitment to the pursuit of science. If medical progress required experiments on animals, then the scientist's cool indifference to vivisecting them signified dedication and self-mastery rather than callousness or cruelty. By the early-twentieth century, vivisection had come to be seen as an indispensable weapon in medicine's unending 'fight' against disease: to be pro-vivisection was to be for science, progress, and the relief of human suffering, while anti-vivisectionists were enemies of science, whose sentimentality and squeamishness were obstacles to be overcome.

To understand the motives of anti-vivisectionists, one must first appreciate the fundamental differences over the nature and goals of science that lay (and perhaps still lie) at the heart of the debate. I have laid particular emphasis on the thesis that vivisection was the expression of a particular view of science—objective, dispassionate, materialistic—which has now become so familiar it is largely taken for granted, but which was contentious in its time. Anti-vivisectionists also regarded themselves as scientific—they tried to test their position through experiments such as anti-vivisection hospitals—but their idea of what constituted scientific sources extended beyond results acquired in the clinic and laboratory to encompass less palpable forms of knowledge. To them, a scientist's duty was not simply to investigate physical phenomena but to seek a deeper appreciation of their significance and, by paying attention to ethical and social considerations, to increase the sum of human wellbeing.

The spiritual revival, and 'new age' utopianism, deeply influenced the anti-vivisection movement. While the most important motives for early anti-vivisectionists were probably their commitment to Christian values and personal virtue, it is striking how many later ones were theosophists, vegetarians, pacifists, and members of various socio-religious orders, unions and groups. For these reformers, vivisection was unscientific because it treated animals as mere matter and ignored the spiritual, aesthetic and moral aspects of life that, though intangible, had to be heeded if humankind's harmony with nature was to be restored. The idea that the vivisectionists' mask of objectivity hid their inability to understand the real consequences of their actions, or even accept their own feelings, persisted in animal welfare writing until at least the late-twentieth century, when Air Chief Marshal Lord Dowding (1882–1970) dismissed them as 'so-called scientists'.[11]

As fascinating and revealing as the history of vivisection and its opponents can (I hope) be, the writings on both sides are voluminous, repetitive, and, for the modern reader, wearisome to plough through. There quite possibly never was a contest in which the disputants failed so comprehensively to grasp one another's point of view. John Simon, speaking at the International Medical Congress in London in 1881, summed it up thus:

> Our own [*i.e.* the experimenters'] verb of life is εργαζεσθαι [to work], not αισθανεσθαι [to perceive]. We have to think of usefulness to man. And to us, according to our standard of right and wrong, perhaps those lackadaisical aesthetics [of anti-vivisectionists] may seem but a feeble form of sensuality.

Of course, not even Mr Simon could entirely numb himself to suffering: he went on to say, without irony, that he found vivisection 'painful'.[12] For all their talk of objectivity, many experimenters found themselves in a similar position: according to Rob Boddice, Victorian physiologists used anaesthetics to spare their own sensibilities, rather than out of consideration for the animals.[13] Fearful of a public outcry, the pioneers of animal experimentation imagined potential anti-vivisectionists around every corner, though, as with most other moral issues then and since, the majority of people remained stubbornly indifferent. It took histrionics to break through their complacency, and both sides obliged. Rather than repeat these recurring diatribes in every chapter, it will suffice to summarise them here.

Vivisectionists, according to their critics, were cruel, sadistic, and even diabolical in their wickedness. There are, of course, no facts to support this; in reading many accounts, I have found nothing to suggest that any British vivisectionist derived perverse pleasure from his work.[14] So far as one can tell, all thought they were doing something that needed to be done, and many genuinely disliked doing it. As for the accusations that they were cold, heartless and indifferent, this they did not dispute, though they wrote cool, objective and scientific instead.

For them, the whole point of the scientific method was that one was supposed to be dispassionate.

In turn, experimentalists dismissed anti-vivisectionists as soft, sentimental, and womanish, accusing them of valuing other animals above their own species and hampering life-saving research because they were too weak to stomach the necessary experiments. They knew, in their hearts, that medical science must progress, but were too cowardly, or hypocritical, to help: no wonder some were also pacifists. Personally, I would consider most of the comments directed at anti-vivisectionists quite flattering if applied to me, and reassuring if applied to the doctor about to treat me. Some anti-vivisectionists acknowledged the same, but the culture of masculinity was strong in medicine and this, along with the high importance accorded to laboratory work, explains the otherwise surprising fact that many members of this quintessentially caring profession were driven to regard sentimentality as a weakness of character.

In Chap. 2, I suggest that the vivisection debate in the early-nineteenth century was more about virtue than utility, and that much of the opposition to vivisection in Britain came from medical men who felt it would bring their profession into disrepute by linking them with cruelty. In contrast to more recent debates, the question of what results vivisection was expected to yield was less important than whether a virtuous person ought to do it. This concern about the vivisector's character was particularly germane to doctors, upon whom there rested a professional obligation to behave with compassion. For the anti-vivisection movement, the spectre of 'human vivisection', their alarmist term for experimentation on hospital patients, proved a good recruiting sergeant, since the poor could be convinced that doctors who vivisected animals might well regard the sufferings of their charity patients with similar indifference. The medical profession's defence was that any 'inhumanity' on the part of doctors who experimented was not a failing of character but a deliberately cultivated cold-bloodedness, which was, as John Hunter famously put it, 'necessary'.[15]

Chapter 3 considers the importance of the theological and metaphysical status of animals for the anti-vivisection movement, particularly the question, now somewhat neglected but frequently raised in the nineteenth century, of whether animals have souls. Changes in praxis in Victorian Britain indicate that popular faith moved ahead of 'official' theology, as people mourned the deaths of animals and speculated on the possibility that humans and non-humans might share a common afterlife. It would be almost a century before most Christian denominations would seriously address this and other issues related to animals, and though many animal welfare projects were initiated with Christian intent, no mainstream church had much to say about the subject, and some defended vivisection.

A perception of greater spiritual kinship between humans and animals arose, I suggest, through the influence of a variety of non-Christian sources, including classical paganism, esoteric philosophy, transcendentalism, and Western interpretations of so-called 'Eastern' religions, the same influences that, channelled through the utopian 'new age' organisations that sprang up around the turn of the nineteenth century, would eventually inspire the environmental movement. The connection between what I shall call 'new age' thinking and anti-vivisection stemmed from a shared rejection of the agenda of scientific materialism in favour of a holistic worldview, in which intuition and feeling were as important as observation and experiment.

Chapter 4 looks at the 'new age' programme through the work of Josiah Oldfield, the founder and medical director of Britain's first vegetarian and anti-vivisection hospitals, and secretary of the Order of the Golden Age—a Christian organisation committed, amongst other things, to health reform and pacifism. Oldfield was a campaigner for fruitarianism and anti-vivisection, and his career provides a case study of how medical opposition to cruelty was linked to aspirations for social and spiritual reform. His anti-vivisection hospital, a converted South London town house with only eleven beds was, as its critics eagerly noted, not intended solely for the benefit of its patients. Indeed, there were more vice-presidents than patients, the object being to attract sufficient public support to prove that an anti-vivisection hospital was a financially viable venture.

Chapter 5 surveys the work of the National Anti-Vivisection Hospital, which opened in 1903 and continued Oldfield's project on a larger scale, challenging the state-sponsored 'vivisecting' hospitals and acting as a showcase for compassionate medicine. While it is likely that no animals were directly saved by the hospital's existence, it did spare the anxieties of the local poor, who feared hospitals that permitted vivisection would not scruple to experiment on their patients. As a propaganda exercise it was a victim of its own success: advertised as the only hospital to which anti-vivisectionists were able to donate with a clear conscience, it was seen as a threat to the funding of other voluntary hospitals, and the King's Fund, Research Defence Society and others instigated a financial boycott that forced it to close.

Chapter 6 looks more closely at the vivisectionists' response, particularly as mediated through the work of the Research Defence Society. Founded in the early-twentieth century by a group of physiologists who advised the government on granting vivisection licenses under the 1876 Cruelty to Animals Act, it had far fewer members, and less funds, than any of the main anti-vivisection groups, but managed to exert disproportionate influence. Through astute political manoeuvring it blocked anti-cruelty legislation in parliament, curtailed advertising by anti-vivisection societies, and arranged for their charitable status to be revoked. Anti-vivisectionists complained that shadowy vested interests were conspiring against them, but it was their own links with fascist groups and their opposition to wartime research and vaccination programs that cost them sympathy and led the government and public to see their activities as unpatriotic and even treasonable.

Chapter 7—After the Second World War, anti-vivisection languished as other untarnished social causes competed for money and support. This was the era of the National Health Service, and government control of medicine was also extended to drug testing. Lethal dose testing in particular led to a large increase in the number of animals used, mostly as part of standard drug testing protocols, which left experimenters no longer personally responsible for deciding what procedures they carried out. The deaths of animals specifically bred for the laboratory, in experiments reported only through the publication of official statistics, lacked the emotive impact of the public vivisections of companion animals that had given the anti-vivisection movement its early impetus. While most of the suffering was now hidden, a few well-publicised animal experiments in high tech contexts, such as the field of space exploration, accustomed the public to equate them with prosperity and progress.

While I have tried to include in this book all the major events in the history of anti-vivisection in Britain from the nineteenth century until the 1960s, I have favoured some aspects that are less well covered in the classic studies by John Vyvyan, Richard French, and Nicolaas Rupke (to which the reader may wish to refer), and have focussed on the underlying ethical concepts and social trends at the expense of providing detailed histories of the major anti-vivisection groups and their leaders, as these are, happily, becoming quite well known.[16] I have not described vivisection experiments in detail, because it is not my intention to write a scientific history of vivisection, nor to turn the reader's stomach. Vyvyan's work has been a particular inspiration, and I have followed his lead in that, while striving to be accurate and to do justice to the arguments on both sides, I have not pretended to be dispassionate about the events l describe. Indeed, it would be wilfully obtuse to write so much about the virtues of sympathy and compassion and then affect indifference to the infliction of so much suffering over so many years.

Notes

1. Throughout the text, 'animals' is used for 'non-human animals', lest repeated iterations of the latter phrase become wearisome.
2. Justin Oakley, 'Virtue ethics and bioethics', in Daniel C. Russell (ed.) *The Cambridge Companion to Virtue Ethics* (Cambridge: Cambridge University Press, 2013), 197–220 (Oakley 2013).
3. Alberto Giubilini and Francesca Minerva, 'After-birth abortion: why should the baby live?', *Journal of Medical Ethics*, *39* (6), 261–263 (Giubilini and Minerva 2013).
4. Julian Savulescu, '"Liberals Are Disgusting": In Defence of the Publication of "After-Birth Abortion"', http://blogs.bmj.com/medical-ethics/2012/02/28/liberals-are-disgusting-in-defence-of-the-publication-of-after-birth-abortion/, viewed 5 March 2016 (Savulescu 2016).
5. The Oxford English Dictionary defines 'humane' as 'such behaviour or disposition towards others as befits a man', the sense in which it is used in this book.
6. Peter Singer, 'Animals and the Value of Life', in Tom Regan (ed.) *Matters of Life and Death: New Introductory Essays in Moral Philosophy* (New York: Random House, 1980), 218–259, on p. 254 (Singer 1980).

7. Tristram Stuart, *The Bloodless Revolution: A Cultural History of Vegetarianism from 1600 to Modern Times* (New York, W.W. Norton, 2007, 71–72 (Stuart 2007).

8. Stanley Godlovitch, Roslind Godlovitch and John Harris (eds) *Animals, Men, and Morals: An Enquiry into the Maltreatment of Non-Humans* (New York: Taplinger, 1972); Peter Singer, *Animal Liberation* (London: Pimlico, 1995), 270 (Godlovitch et al. 1972; Singer 1995).

9. Andrew Linzey, Animal Theology (London: SCM Press, 1994), 20, 41–42 (Linzey 1994).

10. 'The non-alcoholic treatment of disease' *BMJ*, *1* (1878), 830–831 (The Non-alcoholic Treatment of Disease 1878).

11. John Vyvyan, *In Pity and in Anger: A Study of the Use of Animals in Science* (London: Michael Joseph, 1969), 7 (Vyvyan 1969).

12. John Simon, 'An Address delivered at the opening of the Section of Public Medicine', *BMJ*, *2* (1881), 219–223, on p. 223 (Simon 1881).

13. Rob Boddice, 'Species of compassion, aesthetics, anaesthetics, and pain in the physiological laboratory', *Interdisciplinary Studies in the Long Nineteenth Century*, *15* (2012), 19 (Boddice 2012).

14. As there were apparently no female vivisectionists in nineteenth-century Britain, I have employed the male pronoun, as to do otherwise would be to accord women a false equality as ahistorical as it is unflattering.

15. See Ruth Richardson, 'A necessary inhumanity?', *Medical Humanities*, *26* (2001), 104–106 (Richardson 2001).

16. R.D. French, *Antivivisection and Medical Science in Victorian Society* (Princeton, NJ: Princeton University Press, 1975); Nicolaas A. Rupke (ed.), *Vivisection in Historical Perspective* (London: Routledge, 1990) (French 1975; Rupke 1990).

References

Boddice, R. (2012). Species of compassion, aesthetics, anaesthetics, and pain. *Interdisciplinary Studies in the Long Nineteenth Century, 15*, 19.

French, R. D. (1975). *Antivivisection and medical science in victorian society*. Princeton, NJ: Princeton University Press.

Giubilini, A., and Minerva, F. (2013). After-birth abortion: Why should the baby live? *Journal of Medical Ethics, 39*(6), 261–263.

Godlovitch, S., Godlovitch, R., & Harris, J. (Eds.). (1972). *Animals, men, and morals: An enquiry into the maltreatment of non-humans*. New York: Taplinger.

Linzey, A. (1994). *Animal theology.* London: SCM Press.

Oakley, J. (2013). Virtue ethics and bioethics. In D. C. Russell (Ed.), *The Cambridge companion to virtue ethics* (pp. 197–220). Cambridge: Cambridge University Press.

Richardson, R. (2001). A necessary inhumanity? *Medical Humanities, 26,* 104–106.

Rupke, N. A. (Ed.). (1990). *Vivisection in historical perspective.* London: Routledge.

Savulescu, J. (2016). "Liberals Are Disgusting": In Defence of the Publication of "After-Birth Abortion". Retrieved March 5, 2016, from http://blogs.bmj.com/medical-ethics/2012/02/28/liberals-are-disgusting-in-defence-of-the-publication-of-after-birthabortion/.

Simon, J. (1881). An address delivered at the opening of the section of public medicine. *British Medical Journal, 2,* 219–223.

Singer, P. (1980). Animals and the value of life. In T. Regan (Ed.), *Matters of life and death* (pp. 218–259). New York: Random House.

Singer, P. (1995). *Animal liberation.* London: Pimlico.

Stuart, T. (2007). *The bloodless revolution: A cultural history of vegetarianism from 1600 to modern times.* New York: W.W. Norton.

The Non-alcoholic Treatment of Disease. (1878). *BMJ, 1,* 830–831.

Vyvyan, J. (1969). *In pity and in anger.* London: Michael Joseph.

2

Vivisection, Virtue, and the Law in the Nineteenth Century

The history of vivisection is inseparable from that of medical science. Without animal experimentation, the course of medicine would have been radically different (one can admit as much without making any presumption about the validity of animal models). Since the nineteenth century, laboratory experimentation has become the gold standard of academic medicine, shaping not only its approach to solving problems, but also the moral conduct and education of doctors. To experimentalists, it was axiomatic that medical science must be objective, rational, and dispassionate: if its advancement required the infliction of pain on laboratory animals, then it was unprofessional, even unethical, to allow squeamishness or sentiment to get in the way. Thus there arose a tension in medicine between the scientific spirit of cool indifference to suffering and the clinical tradition of compassion and caring. When the Continental fashion for vivisection first touched Britain in the 1820s, many doctors chose to distance themselves from it for the sake of their reputation, and the few who did undertake it felt the need to defend a choice that seemed at odds with the ethos of their profession.

Though 'anti-vivisection' became, in the course of the nineteenth century, so familiar a term of self-description that it would be obtuse to call organized opposition to animal experimentation by any other

© The Author(s) 2017
A.W.H. Bates, *Anti-Vivisection and the Profession of Medicine in Britain*,
The Palgrave Macmillan Animal Ethics Series, DOI 10.1057/978-1-137-55697-4_2

name, its use has perhaps distorted the way that both the individuals and organizations involved have been understood, implying as it does protest, negativity and perhaps even rejection of progress—a campaign by some out of step luddites and radicals to halt the march of science, or to make a heavy-handed moral point about the abuse of power within society. It would, however, be no less apt to view vivisectionists as the protesters and their opponents as the conservative majority. There was never a time in Britain when there were more people active in support of vivisection than against it, and in the nineteenth century the antis raised petitions with hundreds of thousands of signatures, more than for any other cause of the time. A key question for the historian is why, considering the level of popular support and money at their disposal, anti-vivisectionists made so little progress in curbing, still less ending, experiments on animals. The remarkable success of experimentalists in winning over the government, legislature, and universities, and in carrying through their objectives in a nation of reputed animal lovers, which proved critical in shaping the course of medical science and ethics, also calls for an explanation.

During the nineteenth century, the anti-cruelty lobby went from being largely unaware of vivisection to passionately opposing it, largely due to a few high-profile incidents. By the century's end, anti-vivisection had become a humanitarian *cause celebre*, a mainstream issue with great public support and many societies dedicated to it, despite vivisection being responsible for only a tiny fraction of the vast amount of suffering inflicted on animals by human hands. Vivisection was seen as different from other forms of cruelty, such as the mistreatment of farm and draught animals, partly because those responsible were linked with the healing and academic professions, whose morality was supposed to be beyond reproach, and also because it had implications beyond animal welfare: for the way society made ethical choices, for how science should be conducted, and for how humans saw themselves in relation to the rest of creation. Because of the multiplicity of moral problems that vivisection raised, the anti-vivisection movement attracted people with all sorts of religious, political, and social principles to defend, and with such disparate views that, according to Vyvyan, anti-vivisection was often the only thing they had in common.

Legislating Against Cruelty

The resurgence of interest in animal protection in the 1970s prompted historians to revisit its history, and pioneering studies such as those of Richard French set early-nineteenth century anxieties over cruelty to animals in the context of post-Revolutionary concerns about the damage that a culture of violence might do to human society.[1] According to French and to Anita Guerrini, calls for laws to protect animals were primarily a form of social self-defence, requiring the government, with the cooperation of middle class activists, to legislate to control the violent impulses of an underclass who lacked the wherewithal to regulate themselves, and who, if permitted to acquire a taste for blood, might soon become ungovernable. At the other end of the spectrum of anthropocentric concern for mistreated animals there were radicals, particularly feminists and socialists, who saw animals as surrogates for disempowered humans.[2] If one adds to these British doctors who resented the introduction of novel, Continental experimental methods that challenged their tradition of observational bedside medicine, it can be seen that there were plenty of people calling for a ban on vivisection who were concerned with their own interests rather than those of animals.

Walter Bagehot famously said that one cannot make men good by act of parliament, but early-nineteenth century anti-cruelty legislation was an attempt to do just that. The sight of bloodthirsty crowds at cock-fights and bull-baitings was especially disturbing for an urban bourgeoisie still unsettled by the French Revolution. If casual cruelty to animals led to, or at least stoked a propensity for, violence against humans (a link, incidentally, that is now well established[3]), then its elimination would be in society's interest. Parliament, however, was slow to act, partly because its members could not decide if blood sports were incitements to cruelty or safe outlets for high spirits.[4]

Attempts by private members to put legislation through parliament finally met with success in Richard Martin's (1754–1834) Cruel and Improper Treatment of Cattle Act (3 George IV. c. 71), which became law in 1822. Martin—an Irish politician, duelist and gambler on whom George IV bestowed the nickname 'Humanity Dick'—had been elected to the UK parliament at the time of the Act of Union, but his bill was of little interest to most of its members, who greeted it with mocking laughter.[5] That there was enough support for it to be passed at all was mostly due to London's changing demographic: in the crowded capital, the well to do could not avoid witnessing the brutal treatment of draught animals and livestock. Prior to Martin's Act, it had been against the law to harm an animal only if it were someone else's property, an offence equivalent to criminal damage, but the Act criminalized cruelty to one's own 'cattle' (though not domestic animals). Since there was no funding to enforce it, the law's effectiveness depended upon private citizens being willing to report acts of cruelty and prosecute those responsible in the magistrates' courts.[6]

Voluntary associations such as the Society for the Prevention of Cruelty to Animals (SPCA), founded in 1824, were set up to help gather evidence and bring prosecutions.[7] Their founders hoped this would improve the 'moral temper' of the populus,[8] but the legal requirement for an offender to have acted 'wantonly and cruelly' made convictions for cruelty rare, since any rational act, however heartless, was not 'wanton' in a court of law. For example, a man who whipped a goat pulling a cart was found not guilty because he said he had beaten the animal in order 'to make it go'.[9] A far-reaching consequence of this, and subsequent, anti-cruelty legislation was that people focused on staying within the law rather than doing what was right, turning cruelty from a moral to a legal problem. The best that anti-cruelty societies could hope for was that fear of prosecution would make the urban poor, who they considered to be the chief offenders, take greater responsibility for their own actions, and thus 'compel them to think and act like those of a superior class'.[10]

Martin's Act did not mention vivisection, probably because he was unaware of it. It was a sign of their rarity in Britain that even someone as concerned as he about animal welfare does not seem to have known about experiments on animals until 1824, when the French physiologist François Magendie (1783–1855) gave a widely-reported demonstration at London's Windmill Street anatomy school.[11] The published accounts of Magendie's experiment, in which he nailed a greyhound to the dissecting-table before cutting it open, provoked a vociferous anti-French outcry that marked the start of the organized anti-vivisection movement in Britain.[12] Thereafter, Martin spoke out against Magendie in particular and vivisection in general, but his own law was powerless to stop it because scientific experiments were performed in a deliberate, calculated manner and not 'wantonly', and so could not, by definition, be cruel under the law.[13] The need to show they were compliant with anti-cruelty law inclined future experimenters to favour utilitarian arguments, because anticipated benefits to human health provided a clear, rational justification for their work.

Medical Opposition to Vivisection

At the time of Magendie's arrival in England, vivisection was 'proverbially rare'.[14] It has been estimated that in the 1820s fewer than a thousand experiments on animals were performed each year in the whole of the British Empire, and English medical men were said to have a particular 'horror' of them.[15] At this time, the medical 'profession' was a loosely defined group with no common licensing or regulatory body, so there was no 'official' position on animal experimentation, but groups of practitioners, concerned that physiological experiments would give medicine a 'bad name', signed up to anti-vivisection testimonials.[16]

Particularly worrying for medics was the rumour that Magendie found experimentation pleasurable, a shocking claim, though not an unreasonable one, as according to a medical eyewitness who attended his demonstrations in Paris: '…he really likes his business… when loud screams are uttered, he sometimes laughs outright'.[17] Magendie's supporters seem to have been aware of the damage that a reputation for callousness could do to his medical career, since they published a testimonial to his kindly bedside manner, presumably aimed at reassuring patients and others that his heartlessness in the laboratory did not extend into his clinical practice.[18]

In the wake of the Magendie scandal, the secretary of the SPCA seized the opportunity to solicit the support of prominent medical men for 'a board … of the profession, to whom all proposed experiments must be submitted'.[19] The medical dignitaries to whom he wrote all declared themselves against unrestrained vivisection, and a selection of their replies, calling for curbs on animal experiments, was published in the national press.[20] The SPCA wanted a panel of distinguished doctors to decide what experiments should be permitted, but to Lewis Gompertz (1783/1784–1861), who founded the rival Animals' Friend Society (AFS) in 1833 after resigning from the SPCA, this was unacceptable. Gompertz, who rejected scientific 'necessity' as a justification for cruelty, and wanted the wording of Martin's Act changed to read 'wantonly *or* cruelly', had no faith in medical self-regulation. He offered a cash reward for evidence leading to the conviction of surgeons or medical students who 'cut up Dumb Animals Alive',[21] though it was never claimed because, as Thomas Wakley (1795–1862), the outspoken editor of the medical journal the *Lancet*, complacently pointed out, anti-cruelty law did not apply to medical experiments.[22] In a misguided attempt to remedy this, the AFS campaigned for, and got, an extension of Martin's law to include the domestic animals experimenters usually used, but this made no difference as experiments on them would still not be judged 'wanton' by a magistrate.[23]

The Character of the Vivisector

One London medical journal accused doctors who opposed vivisection of trying to 'curry favour' with patients, which suggests that patients were known to prefer doctors who were not vivisectionists.[24] Some people were afraid that the vivisectors' real objective was to experiment on humans, and as charity patients were thought to be their most likely victims, anti-vivisection became a popular cause in poor districts of London.[25] Paying patients had less reason to fear being used as experimental material, but they tended, as patients still do, to choose their doctor on the basis of personal attributes such as compassion, kindness and humanity rather than on purely technical ability, which made anti-vivisection a pragmatic position for medics in private practice to adopt.[26]

There was a longstanding, though apocryphal, tradition that anatomists and butchers were banned from serving on coroners' juries because their trades destroyed their moral competence, and vivisection was presumed to have a similar effect, especially on impressionable young students.[27] According to the Irish physician and naturalist James Lawson Drummond (1783–1853), 'little was to be expected' of medical students who became habituated to vivisection, a concern echoed by the SPCA's president Lord Carnarvon, who claimed that it was because they knew that the vivisector's 'feelings of compassion for suffering [could become] entirely obliterated', that the majority of medics wanted it restricted by law.[28]

There is no official record of how many experiments on animals were performed in Britain before the 1876 Vivisection Act, but few were reported in medical journals (most of the experiments published in British journals were performed in France or Germany) and they do seem to have been truly rare. Concerns within the medical profession were probably instrumental in keeping them so, though the few British doctors who did vivisect met with less criticism in the press than their Continental counterparts, partly because there was an element of nationalism behind protests against French physiologists, but also because the English tended not to carry on their work in public.

Opponents of Continental style vivisection demonstrations feared that these open displays of cruelty by professional people would lead to a general moral decline.[29] A well-known account by French physiologist Claude Bernard (1813–1878) of a Quaker who berated Magendie in his own laboratory provides a concise summary of public concerns:

> Thou performest experiments on living animals. I come to thee to demand of thee by what right thou actest thus and to tell thee that thou must desist from these experiments, because thou hast not the right to cause animals to die or make them suffer, and because thou settest in this way a bad example and also accustomest thyself to cruelty.[30]

A review of the case against vivisection by the AFS in 1833 grouped objections to it under five headings: to inflict pain on animals was a moral failing, it engendered public animosity against scientists, encouraged cruelty towards humans, distracted charitable efforts away from human suffering, and offended God.[31] Significantly, even this most committed of anti-vivisection groups relied on anthropocentric arguments: the suffering experienced by animals and their rights or interests were of lesser importance than the effects of vivisection on the experimenter and on society.[32]

British medical practitioners in search of moral guidelines had no governing body to turn to and no written code of conduct to help them; the word 'ethics' was seldom mentioned in medical writing, and there was no specific legislation governing medical practice until the Medical Act of 1858. There was, however, a widely-accepted code of etiquette and personal morality, which, though it included some rules of conduct specific to doctors, such as not stealing a colleague's patients, was largely that of their social group. A medical man was expected to be, or at least act like, a gentleman, a class whose 'honour and humanity are unimpeachable'.[33] For those to whom gentility did not come by birth or upbringing, guidebooks known as gentlemen's manuals provided instruction on correct manners and behaviour: the proper treatment of animals was neither cruel and heartless nor overly emotional, since brutality on the one hand and sentimentality on the other fell short of gentlemanly standards.[34]

I have argued elsewhere that the ideal of gentlemanly medical conduct was, in modern terms, an expression of virtue ethics, a system first described in Aristotle's *Nicomachean Ethics* and distinguished by its focus on motives and character rather than actions.[35] In the eighteenth century, in what has been termed 'the decline of virtue', character-based ethics began to lose ground to utilitarianism and deontology, but medicine still retained an old-fashioned attachment to the virtues and a strong emphasis on the good character of its practitioners.[36] For example, in the *Fortnightly Review* of (1882), the physician William Benjamin Carpenter (1813–1885) wrote that the morality of a pain-giving act lay not in the act itself (deontological ethics), nor in its result (utilitarianism), but in the motive for the act, a test that was also applied to vivisection.[37]

How patients and professionals interpreted the motives behind vivisection was therefore crucial in shaping their response to it. Not a few thought that vivisectors must be callous—like Magendie, they were thought to ignore, or even laugh at, animals' screams—a particularly damning charge for those who treated patients.[38] As long ago as 1758, Samuel Johnson (1709–1784) had fulminated against vivisectionists in his periodical the *Idler*, writing that '[i]t is time that a universal resentment should arise against these horrid operations, which tend to harden the heart and make the physician more dreadful than the gout or the stone'.[39] A century later, a similar sentiment was being quoted with approval by the *British Medical Journal*: 'Let there be no mistake about it: the man who habituates himself to the shedding of blood, and who is insensible to the sufferings of animals, is led on into the path of baseness'.[40]

Virtue ethics did not, however, offer a decisive argument against vivisection. To shun it, as the controversial and outspoken anatomist Robert Knox (1793–1862) did, might be taken as a sign of 'humanity', but others claimed that vivisectionists were prompted by worthy motives such as the desire to alleviate human suffering and to acquire knowledge.[41] The challenge lay in deciding which personal qualities to favour: what Charles Darwin (1809–1882) called the 'virtue' of 'humanity to the lower animals', or the laudable wish to advance medical learning.[42] Virtue ethics did not offer a glib solution: the ideal medical character was a balanced one, and it was 'proverbial' that medics ought to be neither unduly sentimental, lest squeamishness made them shrink from their work in order to spare their own feelings, nor so insensitive that they became callous.[43]

A degree of fortitude and resolve was expected of all well-bred men, and anti-vivisectionists and others who were thought to be deficient in these manly, Christian virtues were criticised as 'effeminate'. As some seventy percent of anti-vivisectionists were women, some of whom used the abuse of animals as a metaphor for their own perceived vulnerability, the cause itself came to be seen as feminine, and men who took it up were subject to accusations of unmanliness.[44] One critic called anti-vivisectionists 'old ladies of both sexes', the implication being that their opinions were formed by emotion rather than intellect and were therefore out of place in a scientific debate.[45] Experimenters faced the opposite criticism, that they were so dispassionate as to be wanting in normal human feeling. That medical practitioners who made *public* statements about vivisection predominantly opposed it, while those who supported it maintained a low profile, suggests that sensitivity was a quality more attractive to potential patients than fortitude and commitment to medical science.

Bad Science

The often-asked question of why most people who objected to vivi-section nevertheless ate meat (or wore leather, or hunted) shows that experimenting on animals was seen as a separate moral issue from eating them, wearing them or chasing them, a distinction sometimes lost on the more logically minded: George Bernard Shaw (1856–1950) was shocked to find himself sharing a platform with hunters and fur-wearers when he spoke at an anti-vivisection meeting.[46] From a historical perspective we are not concerned with whether theirs was a coherent or defensible position, but with why they thought as they did. One explanation of why vivisection seemed to be of a different order from other cruelties was that those who performed it were neither the ignorant poor nor the feckless rich but scientists and doctors, precisely the sort of educated, professional people from whom society expected exemplary standards of conduct.[47]

Nineteenth century science was as much an attitude of mind as a field of study, an objective discipline where feelings did not intrude, and whose practitioners cultivated detachment and self-control. But their duty to suppress any feelings of compassion whilst working was problematic: quite apart from the difficulty of arguing that it was virtuous to suppress a virtue, how was an individual who steeled himself to perform vivisection for altruistic motives to be distinguished from someone so morally indifferent as to give no thought to the suffering he was about to inflict? According to the President of the British Association for the Advancement of Science, an organization whose remit was to defend experimentation, the fact that vivisectionists were 'men of science' meant that, unlike 'persons in the lower order', there was no question of their being unthinkingly cruel, not least because their experiments were too 'tedious and toilsome' to be performed unthinkingly.[48]

One could, however, be heartless without being reckless, and the argument that vivisectionists could control their finer feelings at will did not convince even some doctors, who thought that anyone prepared to inflict pain on helpless animals must be seriously lacking in emotional sympathy. Dr Robert Hull, writing against vivisection in the *London Medical Gazette*, agreed with the surgeon John Abernethy (1764–1831) that any doctor who vivisected was unfit to attend a family, and the *Protestant Magazine* concurred, printing a 'caution to parents' advising them to shun the services of any practitioner who carried out vivisection.[49] This was not a 'slippery slope' argument, since it presumed that vivisectionists were already deficient in empathy: as Immanuel Kant had written, 'we can judge the heart of a man by his treatment of animals'.[50]

The controversy over vivisection shows many parallels with that over human cadaveric dissection that took place in the early 1830s, to the extent that the arguments for and against both were regarded as interchangeable (the public also assumed that anatomists were all vivisectionists, despite their protests to the contrary).[51] Both practices were criticised not because of the suffering they caused but because of the supposed cold-heartedness and self-indulgence of the perpetrators, and both became the subject of public scandals that gave rise to regulatory legislation: the much-publicized murders for dissection in Edinburgh and London between 1829 and 1831 led to the 1832 Anatomy Act that legalized pauper dissection, while the shocking experiments performed by the French physiologist Éugène Magnan on a visit to Britain in 1874 led to the introduction of licensing for vivisectionists through the 1876 Cruelty to Animals Act. Vivisectionists and anatomists alike defended themselves with utilitarian arguments (vivisecting animals and dissecting cadavers were necessary to train doctors and develop new treatments), and also tried to show that their motives were virtuous. The 'heroic anatomist', who set aside any personal feelings and stoically endured the horrors of the dissecting-room, was a prototype of the 'imperturbable scientist' who when experimenting on animals in the laboratory was 'callous for the sake of what he deemed the greater compassion'.[52]

Believing that vivisection might be performed by virtuous people still did not make it good science. Experimenters claimed their work would 'place the art of healing upon a firmer basis', and they certainly obtained objective, quantitative data from animals that could not easily have been got from humans, but there were plenty of methodological objections.[53] A somewhat facile criticism, dating back to the seventeenth century, was that normal function could not reliably be investigated in living animals because their responses under vivisection did not represent a normal state (facile because no interventional experiment can ever study a truly 'normal' state). In the nineteenth century, however, it was subtly modified: animal experiments were bad science because they were a crude and clumsy attempt to wrest Nature's secrets from her by force, rather than through skillful philosophical enquiry, and so they reflected badly on the judgement and finesse of those who resorted to them.[54] Drummond compared vivisection to judicial torture in that it both yielded untrustworthy information and discredited the inquirer, and Karl Marx thought it of doubtful value, and an affront to 'humanity'.[55]

Another common methodological criticism was that animal physiology differed so greatly from the human that results could not be extrapolated.[56] In theory, this was difficult for experimenters to counter because if animals were physiologically similar to humans, they likely felt as humans did and so it was cruel to make them suffer, while if they lacked human sensitivity to pain, they were significantly different from us physiologically: so an experiment was 'criminal' if an animals' physiology was like ours and 'objectless' if it was not.[57] Though this might seem a powerful argument against experimentation, it was unusual for nineteenth-century vivisectors to concern themselves with animal pain at all: like Darwin, they could accept that '[t]he lower animals, like man, manifestly feel pleasure and pain, happiness and misery' and still consider vivisection justifiable.[58] We will consider in the following chapter whether the categorical difference between humans and animals that left the latter vulnerable to experiment was primarily spiritual rather than physiological.

Necessity and Humanity

Though British medical practitioners generally disliked the showy displays by Continental physiologists that had 'drawn odium' upon their profession, they were more sympathetic to experimentation done by their own countrymen, provided it was 'necessary' to medical progress: William Harvey's (1578–1657) work on the circulation of blood and John Hunter's (1728–1793) on aneurysms were the most often cited examples. Writing in the 1860s, the physician and author Andrew Wynter (1819–1876) declared that Hunter's work alone had been worth 'the destruction of a whole hecatomb of dogs', and though few vivisectionists made discoveries of the same magnitude as Hunter's, they all anticipated benefits to humankind and used this to justify their work.[59] Their self-assessment of utility was, of course, subjective: in his monograph *Vivisection Investigated and Vindicated*, the English physician George Etherington described among the animal experiments he thought medically important one in which it was shown that creosote, when injected into dogs, acted as a poison, precisely the kind of obvious but apparently pointless result that opponents thought constituted a strong argument against such experiments.[60]

In despair of ever bringing a successful prosecution against a vivisectionist, Gompertz complained that their being allowed to justify their own experiments on the basis of predicted benefits rendered the law 'nugatory'. Naturally, everyone anticipated that their own work would yield vital results, and so 'necessity' had become 'the cheat of humanity'.[61] Even Etherington agreed that the law was rendered powerless by the stipulation that the offender must act 'wantonly': 'the worst moral character, never performs an act without thinking upon and having a motive in performing it...'.[62] Throughout the nineteenth century, anti-cruelty groups such as the London Anti-Vivisection Society would continue to complain that medical experiments were being performed 'needlessly, and therefore cruelly', but there was no prospect of a successful prosecution as long as the experimenters themselves were the arbiters of necessity.[63]

British vivisectionists were, however, prepared to accept that many Continental experiments were unnecessary, and they criticized the French, in particular, for an excess of speculative studies and public demonstrations. The relatively few British doctors who did vivisect—most notably Marshall Hall (1790–1857), James Blundell (1791–1878), James Hope (1801–1841) and Charles J.B. Williams (1805–1889)—were prudent and discrete, working privately, publishing in professional journals, and following prearranged lines of investigation. Overall, they were successful in avoiding public scrutiny. Hall, for example, repeated some of Magendie's experiments in the 1820s, including one in which he opened the chest of a dog and then made him vomit, whereon, according to Hall's own account, 'a portion of lung was driven through the thoracic opening with violence and a sort of explosion'.[64] Though he was criticized in the medical press for this work, there was no public outcry, probably because lay people were simply unaware of it.[65] Shortly afterwards, Hall published his own 'objective' criteria for deciding whether experiments on animals were justified: the information sought must not be obtainable by observation alone, the experiment must have a distinct and definite object, it must not be a repeat, it must cause the least possible suffering to the least sentient animal, and must be properly witnessed and recorded.[66] Though they made little impact at the time, his rules would influence the drafting of the Antivivisection Act 40 years later.

The stimulus for legislation, when it came, was not the protests of anti-cruelty campaigners, but the continued animus towards French physiologists on the part of British doctors. When Éugène Magnan visited London in 1874, his medical audience, led by Thomas Jolliffe Tufnell (1819–1885), the President of the Irish College of Surgeons, intervened to stop a particularly unpleasant experiment.[67] Magnan promptly returned to France, forestalling an attempt by the now Royal Society for the Prevention of Cruelty to Animals to prosecute him, but, while the British organisers of his demonstration were acquitted of any wrongdoing, the magistrates made it clear what they thought by refusing to award the defendants' costs.[68] This was the closest a vivisectionist in Britain would ever come to being convicted. The case, along with the publication in 1873 of John Burdon-Sanderson's (1828–1905) *Handbook for the Physiological Laboratory*, a *vade mecum* for the 'beginner' that made scant reference to anaesthesia, raised such concerns that continental-style vivisection might become acceptable in Britain that in 1875 the government set up a Royal Commission on the matter, the result of which would be the 1876 Cruelty to Animals Act.[69]

The Vivisection Act and the Victoria Street Society

The 1876 Act (39 and 40 Vict. c. 77), known as the Vivisection Act, mandated that vivisection be performed only for an original, useful purpose. This put an end to the sort of public demonstrations that

Queen Victoria and many of her subjects so disliked, but left open the private use of animals for medical research and teaching.[70] The word 'wantonly' was dropped from the definition of cruelty but as, in the expert opinion of Mr Justice Day (Sir John Day, 1826–1908, a judge well known for sentencing felons to flogging), 'cruelty must be something which cannot be justified', the legal requirements for a conviction remained substantially unchanged.[71] Ironically, it was said that the Royal Commission was convinced of the need for regulation not by anti-vivisectionists, but by the testimony of experts such as the German bacteriologist Emanuel Klein (1844–1925), whose candid admission that he used anaesthesia only for his own convenience—to make the animals easier to handle—spoke volumes about the difference in outlook between scientists and the public:

> When you say that you use them [anaesthetics] for convenience sake, do you mean that you have no regard at all for the sufferings of the animals?
>
> No regard at all.
>
> You are prepared to establish that as a principle of which you approve?
>
> I think with regard to an experimenter, a man who conducts special research, he has no time, so to speak, for thinking what the animal will feel or suffer.[72]

Like the Anatomy Act, the Vivisection Act was permissive rather than regulatory. It required all vivisectionists to hold a license but these were liberally bestowed: by 1891, 676 people had been granted one, a large proportion of whom were given 'special' certificates dispensing them from the need to use anesthesia. To acquire an ordinary license, one needed the signatures of two referees who were professors of physiology, medicine, anatomy, or a related discipline; this kept animal experiments 'in the family', so to speak—most licensees worked in the 'golden triangle' of London, Oxford and Cambridge, and enjoyed the support of the universities and medical royal colleges—as well as giving recognition and authority to the new discipline of experimental physiology, whose professional body, the Physiological Society, was founded in the same year the Vivisection Act became law.[73]

The surviving correspondence from the London-based Association for the Advancement of Medical Research, which advised the government on licensing, makes no mention of a licence application being refused, and it is quite possible that none ever was: in 1954, when the Home Secretary was asked how many applications had been turned down since 1876, he told the Commons the information was 'not available'.[74] There was no successful prosecution during the 110 years the Act remained in force.[75]

Medical practitioners opposed to vivisection, some of whom were critical of the Act, were not involved in the licensing process, and ordinary doctors were said to be 'afraid' to speak out because the system was now overseen by the leaders of their profession.[76] The Act also excluded the public from any involvement in decision making, and made it difficult for them to find out where experiments were taking place, as the licensees' names and addresses were not published. The reason for keeping their identities and locations secret was not fear that they would be intimidated (direct action against vivisectionists was unheard of), but concern that the information might deter patients and donors, and encourage unwelcome efforts by members of the public to gain admittance to demonstrations.[77]

It would be difficult to overestimate the importance of the rise of experimental physiology following the Vivisection Act in shaping the narrative of 'modern medicine', of which Claude Bernard was, in the 1930s, already being called the 'father'. Bernard's influence on George Hoggan (1837–1891), an English doctor who briefly worked in his laboratory, would, however, give life to a very different project. In 1875, Hoggan published a harrowing account of the sufferings of the dogs that Bernard vivisected, though without mentioning him by name. He concluded: '…having drunk the cup to the dregs, I cry off, and am prepared to see not only science, but even mankind, perish rather than have recourse to such means of saving it'.[78]

Hoggan suggested to the Irish writer and social campaigner Frances Power Cobbe (1822–1904) that they form a society to campaign against animal experiments, and in 1875, with the support of Lord Shaftesbury (Anthony Ashley-Cooper, 7th Earl of Shaftesbury, 1801–1885) and the Archbishop of York, William Thomson (1819–1890), they formed the Society for the Protection of Animals Liable to Vivisection, better known as the Victoria Street Society (VSS); in 1897, it would become the National Anti-Vivisection Society (NAVS), with Shaftesbury as its president. The Society's goals included preventing the kind of extreme experiments for which Bernard was notorious from being sanctioned in Britain and trying to get the 1876 Act repealed.[79] In the opinion of the VSS, the Act had led to more experimentation than if vivisection had remained unlicensed, and allowed experimenters to use their 'professional esprit de corps... to secure for themselves prolonged immunity from state interference with their atrocities'.[80]

Horrible, Brutalising, Unchristianlike

Rather than continuing with futile attempts to prosecute vivisectionists, pragmatic campaigners tried to instil compassion into the young through anti-cruelty clubs such as the Band of Mercy movement.[81] It is apparent from the voluminous and sometimes tedious polemics published at this time that enthusiasts for, and critics of, vivisection were now relying on very different arguments. For vivisectionists, the justification of their experiments was a utilitarian one, since the predicted benefits to medicine outweighed any suffering, and they saw their opponents, as the ageing Darwin penned to *The Times* in 1876, as tender-hearted but profoundly ignorant.[82] For their part, anti-vivisectionists laboured the point that anyone who experimented on living animals was callous and insensitive, character traits typically associated with the unthinking lower classes, and certainly undesirable in a medical practitioner or scientist. By the end of the nineteenth century, the sentiment was common among the public that, as Queen Victoria had put it, animal experimentation was: 'horrible, brutalising, unchristianlike', and 'one of the worst signs of wickedness in human nature'. With its focus on reducing the pain experienced by animals and licensing scientists, the Vivisection Act had done nothing to address fears that vivisection 'saps our moral sense', 'blunts our sympathy', and promotes 'ruthlessness and oppression'.[83]

In fact, only a small minority of doctors was ever actually involved with animal experiments, and most preferred to avoid them. Despite their stereotypical portrayal in literature as callous and undisciplined, medical students generally shunned vivisection, and it was little used in British medical schools, where many of the teachers shared anatomist Josef Hyrtl's (1810–1894) view that anyone who could look calmly on vivisection would not make a good physician.[84] The VSS, claiming that the new cadre of licensed, professional vivisectors would become so indifferent to suffering that experimentation would be 'the simple, natural thing to do to any helpless creature in their hands', stoked fears that it would be extended to human subjects.[85] Of course, patients were not tied to tables and cut up except in the pages of sensational fiction, but there were other ways of experimenting. The microbiologist Robert Koch (1843–1910) actually did experiment on paupers; Louis Pasteur (1822–1895) proposed experimenting on prisoners, and the dermatologist Jonathan Hutchinson (1828–1913) delayed the treatment of a patient with a painful disease the better to demonstrate the signs to his students, all actions, according to the VSS, to which no vivisectionist could logically object.[86] Though doctors who vivisected may not have treated patients themselves, they could still set a bad example to those who did: if even the most distinguished scientists, wrote Lewis Carroll (Charles Dodgson, 1832–1898), were careless of the suffering they caused, 'what will be the temper of mind of the ordinary coarse, rough man… of whom the bulk of the medical profession… is made up?'[87]

Wary of being thought at best heartless and at worst dangerous, experimental physiologists liked to emphasise that their chosen work was disagreeable to them. According to one sympathetic account, the real sacrifices were being made not by the animals but the experimenters: 'we have heard a considerable number of physiologists declare unanimously, that all vivisection tires them exceedingly; sometimes so shatters them, that it requires all their power of will to carry the process through to the accomplishment of the aim…'.[88] This at least indicates they were aware of the importance of character and sensibility in determining how others judged their actions. From a utilitarian perspective, the case for vivisection would have been stronger if, in addition to acquiring knowledge from it, physiologists enjoyed their work rather than enduring it; by stating that they undertook experiments reluctantly and at great emotional cost to themselves, they were defending their personal virtue by taking on the persona of the heroic scientist who suffers emotional difficulty through being obliged to transgress normal moral boundaries for the sake of science.

George Romanes (1848–1894), Darwin's disciple, stressed that students of physiology must be none the less gentlemen because they were men of science, though the attitudes characteristic of genteel conduct could be difficult to square with what went on in the laboratory.[89] Burdon-Sanderson's private admission that 'emotional and sentimental states' such as sympathy were an experimenter's 'greatest enemies' implies a more heartless attitude than that typically expected of a gentleman, though some physiologists may have thought privately what Queen Victoria's physician Sir William Gull (1816–1890) declared openly: that gentlemen-scientists were above the law, and that anti-cruelty legislation was 'for the ignorant, and not for the best people in the country'.[90]

For the antis, Cobbe memorably asked if:

> … advancement of the 'noble science of physiology' is so supreme an object of human effort that the corresponding retreat and disappearance of the sentiments of compassion and sympathy must be accounted as of no consequence in the balance?[91]

How people answered such questions would determine whether they gave their money, and trusted their health, to vivisectionists, and whether they saw the rise of laboratory medicine as a major advance or a wrong turning.

Notes

1. French, *Antivivisection and Medical Science* (French 1975).
2. Anita Guerrini, 'Animal experiments and anti-vivisection debates in the 1820s', in C. Knellwolf and J. Goodall (eds.), *Frankenstein's*

Science: Experimentation and Discovery in Romantic Culture, 1780–1830 (Aldershot: Ashgate, 2008), 71–86; Coral Lansbury, *The Old Brown Dog: Women, Workers, and Vivisection in Edwardian England* (Madison, WI: University of Wisconsin Press, 1985); Mary Ann Elston, 'Women and anti-vivisection in Victorian England, 1870–1900', in Rupke, *Vivisection in Historical Perspective*; Hilda Kean, 'The "smooth, cool men of science:" the feminist and socialist response to vivisection', *History Workshop Journal, 40* (1995), 16–38; Hilda Kean, *Animal Rights: Social and Political Change since 1800* (London: Reaction, 1998); Ian Miller, 'Necessary torture? Vivisection, suffragette force-feeding, and responses to scientific medicine in Britain *c.* 1870–1920', *Journal of the History of Medicine and Allied Sciences, 64* (2009), 333–372 (Guerrini 2008; Lansbury 1985; Elston 1987; Kean 1995, 1998; Miller 2009).

3. Andrew Linzey (ed.), *The Link between Animal Abuse and Human Violence* (Brighton: Sussex Academic Press, Linzey 2009) (Linzey 2009).

4. Charles D. Niven, *History of the Humane Movement* (London: Johnson Publications, 1967), 57–58; Kean, *Animal Rights*, 33–34 (Niven 1967).

5. Niven, *History of the Humane Movement*, 61 (1967).

6. Gordon Hughes and Claire Lawson, 'RSPCA and the criminology of social control', *Crime, Law and Social Change, 55* (2011), 375–389 (Hughes and Lawson 2011).

7. SPCA, *Sixth Report and Proceedings* (London: W. Molineux, 1832) (SPCA 1832).

8. Kean, *Animal Rights*, 36 (1998).

9. Anon., *Animals Friend, 7* (1839), 51.

10. 'Society for the prevention of cruelty to animals', *Evangelical Magazine and Missionary Chronicle, 2* (1824), 357–358.

11. 'On experiments on living animals', *London Medical Gazette, 20* (1837), 804–808.

12. J.M.D. Olmsted, *François Magendie: Pioneer in Experimental Physiology and Scientific Medicine in Nineteenth-Century France* (New York, NY: Schuman, 1944), 137 (Olmsted 1944).

13. 'Vivisection per se cannot be ranked among acts of cruelty': see, 'The ethics of vivisection', *Lancet, 2* (1860), 143–144.

14. George F. Etherington, *Vivisection Investigated and Vindicated* (Edinburgh: P. Richard, 1842), 17 (Etherington 1842).

15. François Magendie (transl. E. Milligan), *An Elementary Compendium of Physiology; for the use of Students* (Philadelphia: James Webster, 1824), 261; Etherington, *Vivisection Investigated*, 80 (Magendie 1824).

16. 'Physiology', *Medico-Chirurgical Review*, *3* (1825), 198–200; Etherington, *Vivisection Investigated*, 111.
17. Olmsted, *François Magendie* (1994), 221–222 (Olmsted 1944).
18. 'Bills to prevent cruelty to animals', *Parliamentary History and Review* (1826), 756–775.
19. David Mushet, *The Wrongs of the Animal World* (London: Hatchard and Son, 1839), 209 (Mushet 1839).
20. 'Society for preventing cruelty to animals', *Morning Chronicle*, June 1825, 3 (Society for preventing cruelty to animals 1825).
21. Lewis Gompertz, *Moral Inquiries on the Situation of Man and Brutes; On the Crime of Committing Cruelty on Brutes, and of Sacrificing Them to the Purposes of Man* (Fontwell: Centaur Press, 1992), 147 (Gompertz 1992).
22. J. Chippendale, 'Experiments on animals', *Lancet*, *1* (1839), 357–358 (Chippendale 1839).
23. 'Amendment of the laws to prevent cruelty to animals', *Animals Friend*, *2* (1834), 11–15. The bill (5 and 6 William IV. c. 59) was passed with the help of Quaker MP Joseph Pease.
24. 'Physiology', *Medico-Chirurgical Review*, *3* (1825), 198–200.
25. 'Correspondence on field sports and surgical experiments on living animals', *Animals Friend*, *6* (1838), 21.
26. Trevor Stammers, 'The NHS—no place for conscience', *Catholic Medical Quarterly*, *63* (2013), 12–14 (Stammers 2013).
27. Lloyd G. Stevenson, 'Religious elements in the background of the British anti-vivisection movement', *Yale Journal of Biology and Medicine*, *29* (1956), 125–157. No such restriction on jurors ever existed: Leigh Hunt, *Essays* (London: Edward Moxon, 1841), 47–48 (Stevenson 1956; Hunt 1841).
28. James Lawson Drummond, 'On humanity', *London Medical Gazette*, *24* (1838/1839), 160–163 (Drummond 1838/1839).
29. A.W. Bates, '"Indecent and demoralising representations": public anatomy museums in mid-Victorian England', *Medical History*, *52* (2008), 1–22 (Bates 2008).
30. Quoted in: Mary T. Phillips and Jeri A. Sechzer, *Animal Research and Ethical Conflict: An Analysis of the Scientific Literature: 1966–1986* (New York: Springer-Verlag, 1989), 7 (Phillips and Sechzer 1989).
31. Anon. 'Five reasons against cruelty to animals', *Animals Friend*, *1* (1833), 14 (Anon 1833).
32. Rob Boddice, *A History of Attitudes and Behaviours toward Animals in Eighteenth- and Nineteenth-Century Britain: Anthropocentrism and*

the Emergence of Animals (Lampeter: Mellen, 2007); Anita Guerrini, 'Animal experiments' (Boddice 2007).

33. 'Surgical experiments on living animals', *Animals Friend, 7* (1839), 61–62.

34. Anon., *The Young Gentleman's Book* (London: Baldwin and Cradock, 1832), 249.

35. A.W. Bates, 'Vivisection, virtue ethics, and the law in 19th-century Britain', *Journal of Animal Ethics, 4* (2014), 30–44 (Bates 2014).

36. Thomas Percival, *Medical Ethics, or a Code of Institutions and Precepts Adapted to the Professional Conduct of Physicians and Surgeons* (Manchester: S. Russell, 1803) (Percival 1803).

37. William Benjamin Carpenter, 'The ethics of vivisection', *Fortnightly Review, 31* (1882), 237–246 (Carpenter 1882).

38. Olmsted, *François Magendie* (1944), 221–222 (Olmsted 1944).

39. Samuel Johnson, *Idler, 17* (1758), 64.

40. Anon., 'Vivisections in France', *BMJ, 2* (1863), 215.

41. 'Dr Knox, of Edinburgh', *Medical Times, 10* (1844), 245–246; William Hamilton Drummond, *The Rights of Animals, and Man's Obligation to treat them with Humanity* (London: John Mardon, 1838), 23, 148. (Knox 1844; Drummond 1838).

42. Charles Darwin, *The Descent of Man* (London: John Murray, 1871), 2 vols, 1, 101 (Darwin 1871).

43. 'Body-snatchers', *Church and State Review*, July 1862, 87; James A. Steintrager, *Cruel Delight: Enlightenment Culture and the Inhuman* (Bloomington, IN: Indiana University Press, 2004), 64, 124 (Steintrager 2004).

44. Andreas-Holger Maehle, 'The ethics of experimenting on animal subjects', in Robert B. Baker and Laurence B. McCullough (eds), *The Cambridge World History of Medical Ethics* (Cambridge, Cambridge University Press, 2014), 552–557 (Maehle 2014).

45. Andrew Wynter, *Subtle Brains and Lissom Fingers: Being Some of the Chisel-Marks of Our Industrial and Scientific Progress, and Other Papers* (London: Robert Hardwicke, 1863), 66 (Wynter 1863).

46. George Bernard Shaw, *Prefaces* (London: Constable, 1934), 257 (Shaw 1934).

47. Hayley Rose Glaholt, 'Vivisection as war: the "moral diseases" of animal experimentation and slavery in British Victorian Quaker pacifist ethics', *Society and Animals, 20* (2012), 154–172 (Glaholt 2012).

48. 'The British Association for the Advancement of Science', *Medical Times and Gazette*, 2 (1863), 258–260.
49. Robert Hull, 'On vivisection', *London Medical Gazette*, 32 (1842/1843), 864. 'A caution to parents', *Protestant Magazine*, 6 (1844), 57–58 (Hull 1842/1843).
50. Tom Regan and Peter Singer (eds.), *Animal Rights and Human Obligations* (Englewood Cliffs, NJ: Prentice-Hall, 1989), 24 (Regan and Singer 1989).
51. Wynter, *Subtle Brains*, 67 (1863); Brayfytte, 'Experimental physiology: what it is, and what it asks', *Animals Guardian*, 2 (1891/1892), 39–42.
52. Simon Chaplin, 'The heroic anatomist: dissection and the stoic ideal' (2008), retrieved from http://www.rcpe.ac.uk/library/listen; A.W. Bates, *The Anatomy of Robert Knox: Murder, Mad Science and Medical Regulation in Nineteenth-Century Edinburgh* (Brighton: Sussex Academic Press, 2010), 21; Boddice, 'Species of compassion' (Bates 2010; Chaplin 2008).
53. 'Vivisection at Alford', *Lancet*, 2 (1860), 395–396.
54. Rupke, *Vivisection in Historical Perspective*, 22.
55. Drummond, *The Rights of Animals*, 145; 'Christianity and its effect upon man's treatment of animals considered', *Church of England Magazine*, 6 (1839), 294–296; K.F.H. Marx (transl. J. Mackness), *The Moral Aspects of Medical Life* (London: John Churchill, 1846), 121 (Marx 1846).
56. 'F. Lallemand, Observations pathologiques propres à éclairer plusieurs points de physiologie' (review), *London Medical and Physical Journal*, 53 (1825), 238–245; Robert Knox, 'Some observations on the structure and physiology of the eye and its appendages', *Lancet*, 1 (1839), 248–251 (Lallemand 1825; Knox 1839).
57. M. Lordat, Mental dynamics in relation to the science of medicine', *Journal of Psychological Medicine and Mental Pathology*, 7 (1854), 252–263 (Lordat 1854).
58. Regan and Singer, *Animal Rights*, 27 (1989).
59. Wynter, *Subtle Brains*, 68 (1863).
60. Etherington, *Vivisection Investigated*.
61. 'Necessity the cheat of humanity', *Animals Friend*, 9 (1841), 16.
62. Etherington, *Vivisection Investigated*, 72.
63. 'Our programme', *Animals Guardian*, 1 (1890), 2.
64. Marshall Hall, 'On the mechanism of the act of vomiting', *Lancet*, 2 (1828), 600–602 (Hall 1828).

65. Diana E. Manuel, 'Marshall Hall (1790–1857): vivisection and the development of experimental physiology', in Rupke, *Vivisection in Historical Perspective*, 78–104 (Manuel 1990).

66. Marshall Hall, *A critical and Experimental Essay on the Circulation of the Blood* (London: R.B. Seeley & W. Burnside, 1831), 2–7 (Hall 1831).

67. 'Prosecution at Norwich', *British Medical Journal*, 2 (1874), 751–754.

68. Sarah Wolfensohn and Maggie Lloyd, *Handbook of Laboratory Animal Management and Welfare* (Oxford: Blackwell, 2003), 8 (Wolfensohn and Lloyd 2003).

69. French, *Antivivisection and Medical Science*, 112–158; M.A. Finn and J.F. Stark, 'Medical science and the Cruelty to Animals Act 1876: a re-examination of anti-vivisectionism in provincial Britain', *Studies in the History and Philosophy of Biology and Biomedical Science*, 49 (2015), 12–23 (Finn and Stark 2015).

70. E.M. Tansey, '"The Queen has been dreadfully shocked:" Aspects of teaching experimental physiology using animals in Britain, 1876–1986', American Journal of Physiology, *274* (1998), S18–S33 (Tansey 1998).

71. G. Candy, 'The legal definition of cruelty in relation to the animal world', *Animals Guardian*, 1 (1890), 5 (Candy 1890).

72. Quoted in Vyvyan, *In Pity*, 87.

73. Finn and Stark, 'Medical science' (2015).

74. Wellcome Library, London (hereinafter Well) SA/RDS A3; 'Do they ever say "no"?' (editorial), *Animals' Defender*, March 1954, 46.

75. Cobbe tried and failed to prosecute David Ferrier in 1881.

76. F.L.O. Morris, *The Curse of Cruelty. A Sermon [on Ps. Xxxvi. 6] Preached in York Minster* (London, England: Elliot Stock, 1886), 80 (Morris 1886).

77. Anon., *Animals Guardian*, 2 (1891/1892), 33.

78. Vyvyan, *In Pity*, 28, 77–78.

79. Amongst other tortures, Bernard employed an 'apparatus for determining the effects of heat upon live animals'. He wrote: 'The animals exhibit a series of symptoms always the same and characteristic. At first the creature is a little agitated. Soon the respiration and circulation are quickened. The animal opens its mouth and breathes hard. Soon it becomes impossible to count its pantings; at last it falls into convulsions, and dies generally uttering a cry': Frances Power Cobbe, *Light in Dark Places* (London: VSS, n.d.), 20.

80. 'Our programme', *Animals Guardian*, 1, (1890), 2; G. Candy, 'Should the vivisection act of 1876 be repealed?', *Animals Guardian, 1*

(1891), 49–50; '"Plain truth" past and present', *Animals Guardian*, *2* (1891/1892), 6–9 (Candy 1890, 1891).

81. Chien-hui Li, 'Mobilizing literature in the animal defense movement in Britain, 1870–1918', *Concentric: Literary and Cultural Studies* 32.1 (2006), 27–55 (Li 2006).

82. Charles Darwin, *Times*, 23 June 1876: Darwin Correspondence Project, Letter no. 10,546, viewed 12 July 2016, http://www.darwinproject. ac.uk/DCP-LETT-10546 (Darwin 1876).

83. Samuel L. MacGregor-Mathers, 'The roots of cruelty', *Animals Guardian*, *1* (1890), 27 (MacGregor-Mathers 1890).

84. 'Hyrtl, prof. of anatomy, Vienna, on vivisection as demonstration to students', *Zoophilist*, *1* (1881/1882), 145 (Hyrtl 1881/1882).

85. Glaholt, 'Vivisection as war'; editorial, *Zoophilist*, *1* (1881/1882), 190 (Glaholt 1881/1882).

86. Anon., *Zoophilist*, *1* (1881/1882), 244; 'Items of interest', *Animals Guardian*, *1* (1890), 34–35 (Anon 1881/1882, 1890).

87. Lewis Carroll, 'Some popular fallacies about vivisection', *Fortnightly Review*, *23* (1875), 847–854 (Carroll 1875).

88. 'Two views of the vivisector', *Zoophilist*, *1* (1881/1882), 194.

89. *Times*, 25 April 1881, 10.

90. Lady Burdon Sanderson, *Sir John Burdon-Sanderson: A Memoir* (Oxford: Clarendon Press, 1911), 157; Parliamentary Papers, *41* (1876), Q5482 (Sanderson 1876).

91. Quoted in Boddice, 'Species of Compassion'.

References

Anon. (1833). Five reasons against cruelty to animals. *Animals' Friend, 1,* 14.

Anon. (1881/1882). *Zoophilist, 1,* 244.

Anon. (1890). Items of interest. *Animals Guardian, 1,* 34–35.

Bates, A. W. (2008). Indecent and demoralising representations: Public anatomy museums in mid–Victorian England. *Medical History, 52,* 1–22.

Bates, A. W. (2010). *The anatomy of Robert Knox: Murder, mad science and medical regulation in nineteenth-century Edinburgh.* Brighton: Sussex Academic Press.

Bates, A. W. (2014). Vivisection, virtue ethics, and the law in nineteenth-century Britain. *Journal of Animal Ethics, 4,* 30–44.

Boddice, R. (2007). *A history of attitudes and behaviours toward animals in eighteenth- and nineteenth-century Britain: Anthropocentrism and the emergence of animals.* Lampeter: Mellen.

Bills to Prevent Cruelty to Animals. (1826). *Parliamentary History and Review,* 756–775.

Candy, G. (1890). The legal definition of cruelty in relation to the animal world. *Animals Guardian, 1,* 5.

Candy, G. (1891). Should the vivisection act of 1876 be repealed? *Animals Guardian, 1,* 49–50.

Candy, G. (1891/1892). "Plain truth" past and present. *Animals Guardian, 2,* 6–9.

Carpenter, W. B. (1882). The ethics of vivisection. *Fortnightly Review, 31,* 237–246.

Carroll, L. (1875). Some popular fallacies about vivisection. *Fortnightly Review, 23,* 847–854.

Chaplin, S. (2008). The heroic anatomist: Dissection and the stoic ideal. Retrieved from http://www.rcpe.ac.uk/library/listen.

Chippendale, J. (1839). Experiments on animals. *Lancet, 1,* 357–358.

Christianity and its Effect Upon Man's Treatment of Animals Considered. (1839). *Church of England Magazine, 6,* 294–296.

Darwin, C. (1871). *The descent of man and selection in relation to sex.* London: John Murray.

Darwin, C. (1876, June 23). Darwin correspondence project. *Times,* Letter no. 10,546. Retrieved July 12, 2016, from http://www.darwinproject.ac.uk/DCP-LETT-10546.

Drummond, J. L. (1838/1839). On humanity. *London Medical Gazette, 24,* 160–163.

Drummond, W. H. (1838). *The rights of animals, and man's obligation to treat them with humanity.* London: John Mardon

Elston, M. A. (1987). Women and antivivisection in Victorian England, 1870–1900. In N. A. Rupke (Ed.), *Vivisection in historical perspective* (pp. 259–273). London: Croom Helm.

Etherington, G. (1842). *Vivisection investigated and vindicated.* Edinburgh: P. Richard.

Finn, M. A., & Stark, J. F. (2015). Medical science and the cruelty to animals act 1876: A re-examination of anti-vivisectionism in provincial Britain. *Studies in the History and Philosophy of Biology and Biomedical Science, 49,* 12–23.

French, R. D. (1975). *Antivivisection and medical science in Victorian society.* Princeton, NJ: Princeton University Press.

Glaholt. (1881/1882). Vivisection as war; editorial. *Zoophilist, 1,* 190.

Glaholt, H. R. (2012). Vivisection as war: The "Moral Diseases" of animal experimentation and slavery in British Victorian Quaker pacifist ethics. *Society and Animals, 20,* 154–172.

Gompertz, L. (1992). *Moral inquiries on the situation of man and brutes...* Fontwell: Centaur Press.

Guerrini, A. (2008). Animal experiments and antivivisection debates in the 1820s. In C. Knellwolf, & J. Goodall (Eds.), *Frankenstein's science: Experimentation and discovery in romantic culture, 1780–1830* (pp. 71–86). Aldershot: Ashgate.

Hall, M. (1828). On the mechanism of the act of vomiting. *Lancet, 2,* 600–602.

Hall, M. (1831). *A critical and experimental essay on the circulation of the blood.* London: R.B. Seeley & W. Burnside.

Hull, R. (1842/1843). On vivisection. *London Medical Gazette, 32,* 864.

Hughes, G., & Lawson, C. (2011). RSPCA and the criminology of social control. *Crime, Law and Social Change, 55,* 375–389.

Hunt, L. (1841). *Essays* (pp. 47–48). London: Edward Moxon.

Hyrtl. (1881/1882). Prof. of anatomy, Vienna, on vivisection as demonstration to students. *Zoophilist, 1,* 145.

Kean, H. (1995). The "smooth, cool men of science:" The feminist and socialist response to vivisection. *History Workshop Journal, 40,* 16–38.

Kean, H. (1998). *Animal rights: Social and political change since 1800.* London: Reaktion.

Knox, R. (1839). Some observations on the structure and physiology of the eye and its appendages. *Lancet, 1,* 248–251.

Knox of Edinburgh. (1844). *Medical Times, 10,* 245–246.

Lallemand, F. (1825). Observations pathologiques propres à éclairer plusieurs points de physiologie (review). *London Medical and Physical Journal, 53,* 238–245

Lansbury, C. (1985). *The old brown dog: Women, workers, and vivisection in Edwardian England.* Madison, WI: University of Wisconsin Press.

Li, C. H. (2006). Mobilizing literature in the animal defense movement in Britain, 1870–1918. *Concentric: Literary and Cultural Studies, 32*(1), 27–55.

Linzey, A. (2009). *The link between animal abuse and human violence.* Brighton: Sussex Academic Press.

Lordat, J. (1854). Mental dynamics in relation to the science of medicine. *Journal of Psychological Medicine and Mental Pathology, 7,* 252–263.

MacGregor-Mathers, S. L. (1890). The roots of cruelty. *Animals Guardian, 1,* 27.

Maehle, A. H. (2014). The ethics of experimenting on animal subjects. In R. B. Baker & L. B. McCullough (Eds.), *The Cambridge world history of medical ethics* (pp. 552–557). Cambridge: Cambridge University Press.

Magendie, F. (1824). *An elementary compendium of physiology; for the use of students* (E. Milligan, Trans.). Philadelphia, PA: James Webster.

Manuel, D. E. (1990). Marshall Hall (1790–1857): Vivisection and the development of experimental physiology. In N. A. Rupke (Ed.), *Vivisection in historical perspective* (pp. 78–104). London: Rutledge.

Marx, K. F. H. (1846). *The moral aspects of medical life* (J. Mackness, Trans.). London: John Churchill.

Miller, I. (2009). Necessary torture? Vivisection, suffragette force-feeding, and responses to scientific medicine in Britain c. 1870–1920. *Journal of the History of Medicine and Allied Sciences, 64,* 333–372.

Morris, F. O. (1886). *The curse of cruelty. A sermon [on Ps. xxxvi. 6] preached in York Minster.* London: Elliot Stock.

Mushet, D. (1839). *The wrongs of the animal world.* London: Hatchard and Son.

Niven, C. D. (1967). *History of the humane movement.* London: Johnson Publications.

Olmsted, J. M. D. (1944). *François magendie.* New York: Schuman.

Percival, T. (1803). *Medical ethics, or a code of institutions and precepts adapted to the professional conduct of physicians and surgeons.* Manchester: S. Russell.

Phillips, M. T., & Sechzer, J. A. (1989). *Animal research and ethical conflict: An analysis of the scientific literature: 1966–1986* (p. 7). New York: Springer.

Regan, T., & Singer, P. (Eds.). (1989). *Animal rights and human obligations.* Englewood Cliffs, NJ: Prentice-Hall.

Sanderson L. B. (1911). *Sir John Burdon-Sanderson: A Memoir.* Oxford: Clarendon Press.

Shaw, G. B. (1934). *Prefaces* (p. 257). London: Constable.

Society for preventing cruelty to animals. (1825, June). *Morning Chronicle,* 3.

SPCA. (1832). *Sixth Report and Proceedings.* London: W. Molineux.

Stammers, T. (2013). The NHS—No place for conscience. *Catholic Medical Quarterly, 63,* 12–14.

Steintrager, J. A. (2004). *Cruel delight: Enlightenment culture and the inhuman.* Bloomington, IN: Indiana University Press.

Stevenson, L. G. (1956). Religious elements in the background of the British anti-vivisection movement. *Yale Journal of Biology and Medicine, 29,* 125–157.

Tansey, E. M. (1998). The queen has been dreadfully shocked: Aspects of teaching experimental physiology using animals in Britain, 1876–1986. *American Journal of Physiology, 274,* S18–33.

Wolfensohn, S., & Lloyd, M. (2003). *Handbook of laboratory animal management and welfare.* Oxford: Blackwell.

Wynter, A. (1863). *Subtle brains and lissom fingers: Being some of the chisel-marks of our industrial and scientific progress, and other papers* (p. 66). London: Robert Hardwicke.

3

Have Animals Souls? The Late-Nineteenth Century Spiritual Revival and Animal Welfare

Although many of the nineteenth-century arguments against vivisection were based on its supposed adverse effects on those who performed or witnessed it, the status of its animal subjects was not inconsequential. One could not be cruel or heartless to a Cartesian automaton that lacked feeling, and perhaps not to animals that had, as some Christians claimed, been put on earth solely to provide for human needs. The art critic and social reformer John Ruskin (1819–1900), addressing the Oxford branch of the Victoria Street Society in 1884, said that: 'It is not the question whether animals have a right to this or that in the inferiority they are placed into mankind, it is a question of what relation they have to God…'.[1] To see animals from a divine perspective, it was necessary to decide whether they possessed rational souls, and what happened to those souls after death.

For most of the nineteenth century, the idea that animals might have afterlives was a decidedly unchristian one. The epitaphist of Lord Byron's dog Boatswain (d. 1808) derided the sort of Christians who disapproved of memorialising a dead dog for trying to keep 'a sole exclusive heaven' for themselves. Almost a century later, when the following lines in memory of Rocket the hunting dog were published, the poetic

© The Author(s) 2017
A.W.H. Bates, *Anti-Vivisection and the Profession of Medicine in Britain*,
The Palgrave Macmillan Animal Ethics Series, DOI 10.1057/978-1-137-55697-4_3

conceit that dogs would be reunited with their keepers in paradise still had a decidedly heathen ring to it:

> Is a man a hopeless heathen if he dreams of one fair day
>   When, with spirit free from shadows grey and cold.
> He may wander through the heather in the 'unknown far away',
>   With his good old dog before him as of old?[2]

What became of one's canine companion after death was not a trivial matter; for some Christians, the idea that animals' souls could exist apart from their bodies seemed 'absurd in the extreme' or even 'dangerous'[3]: the divine spark of immortality was the one incontrovertible barrier between humans and other animals, however many biological resemblances scientists might go on to discover.

It is sometimes claimed that science, and Darwinism in particular, improved the lot of animals by replacing the traditional Christian model of a static created order with humans at its earthly summit (just below the angels) with a dynamic model in which higher forms were continually evolving from lower.[4] To put it crassly, people were less likely to ill-treat animals to which they were distantly related. Darwinism certainly made many people think about their kinship with animals, although the idea of a serial affinity between different species (the 'ladder of creation' or 'great chain of being'), and even of species change itself, had been current long before the publication of *The Origin of Species* in (1859).[5] For anatomists, the human–animal boundary had been blurred since at least a century earlier, when the great taxonomist Carl Linnaeus (1707–1778) wrote that he was unable to discover 'the difference between man and the orangoutang, although all of my attention was brought to bear on this point'.[6] By the 1840s, there seemed no prospect of anatomists finding any structure that would categorically distinguish humans from apes in terms of morphology: the popular press responded with sensational tales of ape–human hybrids and mocked the 'siantificle' vogue for dissecting monkeys 'to see … whether like our own specius inside as well as out'.[7]

The most obvious distinction between apes and humans in the nineteenth century was the Christian claim that humans alone had souls made in the image of their Creator, and even this was being challenged, in a religious and philosophical setting rather than a scientific one, through ideas gleaned from classical paganism, Hinduism and transcendentalism. This nineteenth-century re-evaluation of the spiritual status of animals would have profound consequences for animal welfare, by bringing to the theological debate, as evolution did to the scientific one, a changed understanding of the relationship between humans and animals.

It would be a mistake, as some freethinkers did (and some still do), to blame cruelty to animals on the low status accorded them in the Judaeo-Christian tradition before Darwinists, orientalists and humanists managed to knock humans off their pedestal: Christian anti-cruelty campaigners were instrumental in giving Britain the most comprehensive animal protection laws in Europe, which they did for the most part without questioning their God-given dominion over the animals they were protecting.[8] As Coral Lansbury (1929–1991) wrote in *The Old Brown Dog* 'the debate between Singer and [Tom] Regan over the moral status of animals would have bemused the Victorians...'[9]: what mattered to them were the moral consequences of inflicting pain on creatures inferior to themselves. Though they did not owe animals a duty of care, they were bound to pity them, and to avoid any imputation of callousness. Rod Preece comments that Christians were more concerned for animals than were Darwinians, and while it might be more accurate to say they were concerned about the dangers to society of allowing cruelty to animals to go unchecked, they turned out, nonetheless, to be the nineteenth-century laboratory animal's best friends.[10]

Christians and Anti-Vivisection in the Nineteenth Century

Most of the groups active in animal welfare, from the Society for the Suppression of Vice, with its emphasis on saving the working classes from being demoralized by alcoholic drink and cruel sports, through the SPCA and its drive to civilise manners, to the VSS with its ethos of compassion, saw themselves as doing the Lord's work.[11] The SPCA, for example, declared that its programme was 'entirely based on the Christian faith', and they denounced vivisection as 'unchristian'.[12] Insofar as the practice of vivisection repudiated the Christian virtues of mercy and compassion, anti-cruelty campaigners saw it as 'evil', 'fiendish', 'blasphemous', and even 'Satanic'.[13] Furthermore, unlike other cruelties such as hunting or meat eating, it was performed by an educated élite, and there was a risk this would lead those less principled to think it was acceptable to be 'cruel' and 'inhumane' out of expediency, so spreading throughout society a heartlessness that was fundamentally 'unchristian'.[14]

Lord Shaftesbury, arguing in 1879 for a total ban on vivisection, said that, for the sake of one's soul, it would be better to be the vivisected than the vivisector, an attitude that, like Shaftesbury himself, exemplified Christian compassion (the famous memorial to him in Piccadilly represents the Angel of Christian Charity).[15] Caring for animals out of Christian charity had the practical advantages that the intellectual or spiritual status of the animals (so long as they were sentient) was of little consequence, while as a motive for action it was readily comprehensible to most people. Defenders of vivisection might dismiss what they derisively termed a 'Brahminical' love for one's fellow creatures as un-British, sentimental, and heterodox, but it was difficult for them to say the same about mercy and compassion towards the weak.[16] The title of the VSS's journal the *Zoophilist* betokened a love of animals, but the Society declared that the main inspiration for its work was 'a conviction that the spread of mercy was the great cause of civilization'.[17]

Cardinal Henry Manning's (1808–1892) outspoken opposition to vivisection, conspicuous among a general Catholic indifference to animals, also appealed to the most basic of Christian virtues:

Vivisection is a detestable practice…. Nothing can justify, no claim of science, no conjectural result, no hope for discovery, such horrors as these. Also, it must be remembered that whereas these torments, refined and indescribable, are certain, the result is altogether conjectural—everything about the result is uncertain, but the certain infraction of the first laws of mercy and humanity.[18]

The RSPCA, failing to appreciate that Manning's views were not representative of Rome, tried unsuccessfully to get anti-vivisection adopted as official Catholic policy, but Pius IX supposedly rejected plans for an office for the Protection of Animals on the grounds that it would send a misleading message that animals had rights.[19] Anglicans were rather more sympathetic to the cause: the Archbishops of York and Dublin signed the 1875 'Memorial Against Vivisection', and though the Church of England remained officially non-committal, by the end of the century some four thousand of its clergy had declared their disapproval of it.[20]

Very few Christians justified their opposition to vivisection by appealing to the unconventional possibility that animals, like humans, had souls that survived death, though there were exceptions, such as Robert Hull, who remarked that '[t]he vivisectors cannot, of course, enter into the depths of that well-grounded suspicion, that there may be a future existence for the brute creation'. For Hull, who hoped that vivisectors would, in some future existence, meet with recompense from those they had tormented, no one who believed in animal afterlives could possibly experiment on them.[21] Robert Browning's (1812–1889) poem *Tray*, published in 1879, imagined the nightmarish possibility of vivisectors deliberately setting out to study the soul of a dog, but this was deliberate exaggeration to shock the reader (Browning was a vice-president of the VSS); no real-life vivisectionist mentioned animals' souls, and some felt vindicated by the conventional Christian teaching that animals had no existence beyond their earthly lives.

The physiologist James Blundell, for example, defended his use of animals in research by claiming that, since an animal's death was an eternal sleep, it was less grave to kill an animal than a human.[22] His reasoning is not entirely clear, but seems to have been based on the presumption that killing becomes murder only if the victim has a soul: in the Old Testament the blood of humans, not animals, cries to heaven for vengeance, as do the souls of martyrs in the Book of Revelation.[23] Others, however, saw the lack of a future life for animals as all the more reason to be compassionate towards them in this one; according to James Lawson Drummond: '[the brute] has no heaven to look to, no bright anticipation of a period when misery shall cease.... Its life is

its little all', an allusion to these lines by the humanitarian poet Anna
Laetitia Barbauld (1743–1825):

> Or, if this transient gleam of day
>   Be *all* of life we share,
> Let pity plead within thy breast,
>   That little *all* to spare.[24]

Animal Afterlives

The question of whether animals had souls was in one sense trivial: an animal, as the name implied, possessed what is known in the
Aristotelian tradition as a vital soul (*anima*). The Cartesian notion of
animals as automata had little currency outside philosophy schools,
and no British vivisector ever adopted this position: they sometimes
argued that animals did not feel pain in the context of a particular
experiment, but none claimed they were incapable of feeling at all.[25]
Indeed, it would have been difficult for a physiologist to make such
a claim, because the validity of experiments on animals depended on
their anatomy and physiology being similar to our own: nervous sys-
tems organized and functioning like ours could scarcely be found in
animals incapable of feeling the pain they so evidently reacted to.[26]
The real question was not whether animals had souls, but how closely
comparable they were to the souls of humans. Were they rational? Did
they experience emotions? And did they, as the poets fancied, share the
promise of immortality?

The Christian doctrine of the immortality of the human soul had its
roots in the thirteenth century, when Thomas Aquinas had modified
Aristotle's position that human beings were a composite of 'form' (*i.e.*
soul) and 'matter' by adding that the human soul was incorruptible and
persisted after bodily death.[27]Aquinas thus brought Aristotle into line
with the Christian promise of eternal life, though at the cost of leaving
disembodied souls, unable to act or experience, in a kind of intellectual
limbo until the resurrection.[28] These incorruptible souls were unique to

humans: other animals possessed 'sensitive' souls but not rational ones, and were blessed with neither reason nor immortality.[29]

In Britain, the Thomist position that the souls of animals perish at death went largely unchallenged by the reformed churches, though the less contentious issue of their rationality was up for debate. In the seventeenth century, Lord Chief Justice Sir Matthew Hale (1609–1676), an advocate of responsible treatment of animals based on a model of stewardship rather than dominion, attributed to them the faculties of memory, reason and imagination (which he called 'phantasies'), but still denied them immortality—the souls of even 'perfect brutes' would die with them.[30] In the eighteenth century, the never easily defensible position that only humans could reason was assailed from both sides: pigs, horses and dogs could apparently be taught to perform calculations and use language, while feral children, brought up without human society, seemed to lack these capabilities.[31] By the nineteenth century, animals' ability to reason was widely accepted, and it was commonplace for magazines and periodicals to entertain their readers with remarkable accounts of animal sagacity.

Although no major Christian church expressed an official view, many individual clergy were happy to admit that animals had rational souls: according to the evangelical missionary Daniel Tyerman (1773–1828), '[t]o deny that brute animals have souls, is virtually to allow that matter can think; and to put an argument into the mouths of materialists that it will not be easy to rescue from them'.[32] For the Catholic Church, Fr John Worthy, a priest in Liverpool, wrote that the rationality of animals was evident from their actions: even bees, Fr Worthy claimed, were intelligent and acted on reason as well as instinct, and many other species appeared from their actions to be 'highly gifted'. Worthy collected numerous accounts from the press in which animals seemed to display social traits such as kindness, gratitude, and affection, or to use imagination and language. However, despite his obvious admiration for these animals' abilities, and his credulity with regard to some rather farfetched tales, Worthy apologised to any of his readers who thought he had 'lowered the dignity of man's soul and reason, by representing the souls and reason of animals as having any degree whatever of similitude with man...' There was, he concluded, an absolute difference between the souls of animals and humans, namely that only the latter survived death.[33]

Most protestants concurred, taking the Biblical reference to mankind having been made 'in the image of God' to mean that the human soul was uniquely able to exist apart from the body. The majority of British divines accepted this position, though there were, as Preece has shown, not a few distinguished exceptions, including the Cambridge Platonist Henry More (1614–1687), the Quaker George Fox (1624–1691), the Civil War pamphleteer Richard Overton and the founder of Methodism John Wesley (1703–1791), all of whom entertained the idea that animals' souls persisted after bodily death.[34] A few Anglican theologians such as Bishop Joseph Butler (1692–1752) can be added to this list, but animal immortality remained largely a nonconformist position.[35]

In the secular literature, however, there was free speculation that companion animals would have a share in the afterlife. Poets who put forward the idea found a ready audience: indeed, so many toyed with it that the anthologist J. Earl Clauson could devote the whole concluding section of his *Dog's Book of Verse* to 'The Dog's Hereafter'.[36] Of course, this poetical vogue for animal immortality was rooted in sentiment rather than solid theological opinions—one might indeed dismiss it as whimsical, a common critical verdict on poems about animals—but it does suggest there was a mood of popular dissent from the 'official' doctrine of an exclusively human afterlife.

Transmigration

One non-Christian path that the souls of animals might follow after death was familiar to anyone versed in the classics. Usually attributed to Pythagoras and his school, the theory of transmigration of souls, or metempsychosis, postulated that the soul or mind was able to survive periods of incorporeal existence between successive incarnations in humans and animals. Though British classicists had 'flirted' with Pythagoreanism since the 1600s, it did not come to general notice until the mid-eighteenth century, when the surgeon and Orientalist John Zephaniah Holwell (1711–1798) published some notes on the subject along with his sensational best-selling account of the Black Hole of Calcutta, of which he was a survivor.[37]

Transmigration seems first to have been used in print as an argument against cruelty to animals in Barbauld's *The Mouse's Petition*:

> If mind, as ancient sages taught,
>   A never dying flame,
> Still shifts thro' matter's varying forms,
>   In every form the same,
> Beware, lest in the worm you crush
>   A brother's soul you find;
> And tremble lest thy luckless hand
>   Dislodge a kindred mind.[38]

Twenty years later, transmigration featured prominently in Thomas Taylor's seminal but idiosyncratic *A Vindication of the Rights of Brutes*, published anonymously in 1792. Despite being one of the earliest contributions to animal rights theory, Taylor's pamphlet is now seldom quoted, in part because his claim that 'brutes' had rights was deliberately hyperbolic, but mostly because the principal object of his writing, as the title suggests, was to satirise Mary Wollstonecraft's (1759–1797) *A Vindication of the Rights of Women* by showing that a parallel argument could be made for the rights of animals.[39] Taylor found Wollstonecraft's proto-feminism ridiculous: he did not acknowledge the rights of men, women, or animals, though he firmly believed that animals were capable of reason and intelligence, which he thought was obvious from their behaviour, and in particular from their capacity to communicate intelligently with one other. From their ability to reason, Taylor concluded that animals had feelings, rejecting Jeremy Bentham's (1748–1832) argument that reason and feeling were distinct, and asserting that 'sense cannot at all operate without intelligence'.[40]

Despite his obvious appreciation of the intellectual abilities of animals, Taylor insisted that compassionate treatment was not their right but a voluntary expression of human virtue, though his work may have inspired subsequent calls for animals to be granted legal rights. When the Lord Chancellor, Lord Erskine (1750–1823) unsuccessfully introduced a Cruelty to Animals bill in the House of Lords in 1809, he complained that: 'Animals are considered as property only: to destroy or to abuse them, from malice to the proprietor, or with an intention injurious to his interest in them, is criminal; but the animals themselves are without protection; the law regards them not substantively; they have no rights!'[41]

Taylor's pseudo argument for the rights of brutes drew not only on their intellectual capacities but also on various traditions concerning the transmigration of souls. A distinguished translator of Plato and Aristotle, Taylor supplemented his classical sources with examples of transmigration collected from non-European traditions, including those of ancient Egypt, Persia and India. Though his writings give the impression he was more attuned to Hellenistic philosophy than Christianity, he did not profess transmigration as a personal belief, but treated the traditions as '[f]ables [which] indicate that brute animals accord with mankind in the nature of the soul'. In other words, the fact that learned people from so many different cultures accepted transmigration revealed a widespread belief that humans and animals were animated by souls of a similar kind.

Pythagoreanism proved to be an inspiration for two influential nineteenth-century anti-cruelty campaigners: Lewis Gompertz and Thomas Forster (1789–1860). Gompertz, who we encountered in the previous chapter, had been secretary of the SPCA until forced to resign in 1833, probably because, as a Jew and a Pythagorean, he did not fit in with the committee's Christian ethos. After a period running the rival, more radical, Animals' Friend Society, he was readmitted after protesting his 'innocence' of Pythagoreanism, but any change of heart in this regard must have been temporary, since in 1852 he wrote in *Fragments in Defence of Animals* that the souls of animals continued to exist after death in a state of limbo, without thought or feeling, until they were united with a new body. Thus, it was possible to be reincarnated, perhaps as a different species, without having any memory of one's previous lives.[42]

Gompertz's interest in Pythagoreanism was shared by his correspondent and fellow member of the AFS, Thomas Forster. A medical practitioner and convert to Roman Catholicism, Forster's eclectic interests included phrenology, vegetarianism, and 'oriental' philosophy, including 'the holy doctrine of Pythagoras and the Indian school, which ascribes to every living creature an eternal existence'.[43] Transmigration was an important part of Forster's idiosyncratic theodicy: 'Metempsychosis implies the future life and everlasting happiness of all living creatures, we must observe that there is plenty of room in this wide universe for all of them... further, without admitting that Animals will live hereafter, we could not reconcile the universal suffering of the brute Creation with the Divine Goodness'. Forster's mix of Catholic purgatory and Pythagorean rebirth allowed cruelty to be punished and suffering recompensed: those who ill-treated animals would find themselves reincarnated in animal bodies, where they would experience for themselves the sufferings they had once meted out, while the merciful would receive 'some light purgatory in the body of some fortunate and beautiful bird or beast'.[44]

Transcendentalism

Apart from evangelicals, who saw medicine as a Christian vocation, and some Anglican Tories among the profession's leaders, medical practitioners had something of a reputation for scepticism and worldliness. Unlike the universities, medical schools did not require their students to profess the Christian faith in order to matriculate, and a lack of pastoral supervision, combined with the materialistic focus of their training, was thought to incline those whose faith was already weak towards atheism.[45] Medical students were certainly encouraged to examine the relationship between humans and animals with a critical eye: comparative anatomy was a key part of their studies, and the problem of why many species, including apes and humans, had similar body plans was, in pre-Darwinian times, accorded high importance. In the 1830s, students frustrated by Professor Granville Sharp Pattison's (1791–1851) 'total ignorance of and disgusting indifference to new anatomical views and researches' forced him from his post at University College London.[46] The students were not, of course, motivated solely by their scientific curiosity: excited by the July Revolution in France, they wanted to hear the radical new Continental ideas on species change and extinction, known as transcendental anatomy, which seemed to carry a particular resonance in those politically unsettling times.

Transcendentalism, also known as philosophical anatomy, was, in its biological sense, a holistic theory of the interconnectedness of all living things that had its roots in German *Naturphilosophie*, which was in turn based on Goethe's concept of nature as a 'vast musical symposium'. It was introduced into Britain by a small number of influential anatomy teachers, who included the surgeon Joseph Henry Green (1791–1863, a friend of Samuel Taylor Coleridge) and the anatomists Robert Knox, Robert Grant (1793–1874) and Richard Owen (1804–1892). Transcendentalism's appeal to students lay in its potential to transform the rather obscure field of comparative anatomy by supplying a coherent, universal theory that would not only account for species change, but also provide a model in nature for abrupt social changes and political revolutions. This potential to upset the status quo gave transcendentalism a radical appeal that ensured its popularity with undergraduates.

Transcendentalism may be defined (not an easy task) as an attempt to discover, through observation and deduction, the fundamental laws and patterns that govern the dynamic, self-organising processes of nature. It thus resembles the Platonic theory of forms in that generalized patterns or archetypes may be deduced from the appearance of objects in the natural world.[47] For example, that most eloquent, and effusive, of transcendentalists, Robert Knox, wrote of the vertebra as: 'the type of all vertebrate animals, of the entire skeleton ... of the organic world .... It possesses the form of the primitive cell; of the sphere; of the universe'.[48]

Critics found this sort of thing vague and mystical, but to Knox, one of the finest comparative anatomists of his day, transcendentalism had the potential to revolutionize his discipline. Anatomists were no longer confined to *describing* morphology, but could speculate on its phylogenetic and even social significance. For example, from observing, measuring, and dissecting human bodies, a set of ideal proportions could be derived, which corresponded to those seen in classical Greek statues such as those of Apollo and Venus.[49] According to Knox, it was no coincidence that the ideal form discovered through modern anatomical studies was the same as that created by ancient Greek sculptors, who had arrived at it intuitively through their refined appreciation of beauty. Transcendentalists thought it legitimate to employ the aesthetic sense in scientific study; for example, those individuals that most closely resembled the ideal type of a species would be considered by a practised observer to be the most beautiful. The transcendental method involved a combination of detailed observation and intuition: knowledge of the structure of different species could only come through careful dissection, but intuition was required to discern the unifying pattern of which each was a variation.

Using transcendental methods, Knox developed a complex theory of evolution, according to which new species arose through differential development (what we might now call mutations) of a common embryo. According to this, pre-Darwinian, proposal, the pattern of every potential species, including humans, was inherent in the multipotent embryo, from which new forms ('hopeful monsters') were constantly being generated, though they would flourish only if external conditions happened to be favourable.[50] As one of the French founders of the movement, Étienne Geoffroy Saint-Hilaire (1772–1844) put it, 'philosophically speaking', there was 'but a single animal', or, in the words of the literary transcendentalist Ralph Waldo Emerson (1803–1882), 'Each creature is only a modification of the other; the likeness in them is more than the difference, and their radical law is one and the same'.[51]

The effect of British transcendentalism on biological thinking was complex and has yet to be fully explored by historians of science. From the perspective of vivisection, however, it proved a deterrent. Firstly, the transcendental method was essentially observational: careful dissection of animal and human bodies was preferred to vivisecting the living; Knox refused to allow any vivisection in his anatomy schools, a pragmatic as well as a humane attitude because students reluctant to dissect living animals would be attracted to transcendentalism as 'a substitute for vivisection'.[52] Secondly, transcendentalism called for a subjective response to nature: an appreciation of beauty helped students to discern the ideal types that lay behind the imperfect forms they encountered, so they needed to cultivate their feelings rather than suppressing them. Thirdly, the transcendental teachings that all species, animal and human, were derived from a common embryo, that all were equally well suited for the environment in which they lived, and that all might become extinct if conditions changed, underlined the essential unity, and transience, of all creatures. Humans were not lords of creation but part of an interdependent, self-sustaining biological system whose life force could be conceptualised as a collective soul, *anima mundi*.[53]

Much of transcendentalism's wider appeal was due to its being not only descriptive of how nature was organised but also prescriptive of the proper way to live. If nature sanctioned abrupt changes (which was how transcendentalists thought new species evolved), then human revolutions might be part of the natural order, and if that order was, as Goethe had expressed it, part of a vast symphony of nature, then humans ought, as far as possible, to live in harmony with it. Transcendental notions of the harmonies of nature were conducive to a philosophy of 'nature mysticism' or pantheism, the followers of which, as Lloyd G. Stevenson (1918–1988) observed, tended to be 'on the side of the animals'.[54]

Though there was never a prominent transcendentalist movement in Britain like that which flourished in New England around Harvard and the Unitarians, the themes of living in 'harmony with nature' and of animals as our 'brothers' did began to appear in British letters from the late 1830s, the most celebrated writers to show a transcendental influence being Samuel Taylor Coleridge (1772–1834) and William Wordsworth (1770–1850). Its appeal was particularly strong for romantics, who sought an escape from the cruelties of metropolitan living in an idealized pastoralism in which animals were helpers, friends and companions.[55]

Transcendentalism in Britain effectively ended as a scientific movement with the publication of *The Origin of Species*, as Darwin's elegantly simple proposal of natural selection made transcendentalism's complex, esoteric explanations of why so many different creatures showed such striking anatomical parallels seem redundant. In contrast to revolutionary transcendentalism, Darwin's modest proposal that change could only occur by gradual small steps was considerably more congenial to the Victorian political establishment.[56] There remained, however, an undercurrent of transcendentalism in biological thought, difficult to trace and sometimes surfacing in unexpected places: the concept of natural harmonies, for example, may have been the inspiration for what are now known as ecosystems. Encounters with transcendentalism during medical training may also have motivated some young British doctors to look more closely into the spiritual aspects of biological phenomena, and to become more open to accepting intuition and emotions as evidence, an approach that would find expression in the *fin-de-siècle* spiritual revival. The theosophists, occultists, new age thinkers and others, within and outside medicine, who became involved in this idealistic movement to unite science and spirituality, some of whom we will encounter in subsequent chapters, might be seen as continuing what was begun by the transcendental anatomists.

Animals' Souls and Anti-Vivisection in the Nineteenth Century and After

The waning of the influence of Christianity on the anti-cruelty movement in the twentieth century coincided with a greater focus on animals themselves. Despite accepting, and even admiring, animals' rationality and learning, the mainstream Christian denominations never bridged the gulf between animals and humans: only the latter were made in God's image and could expect a place in the hereafter. By the late-nineteenth century, however, the British people had already granted them one. Those middle class (for the most part) late Victorians who looked forward to being reunited with their companion animals in the world to come, and mourned their passing in this, probably felt neither hopeless nor heathen. They developed rituals, resembling human funeral practices, to mark the passing of their animal companions: post-mortem photographs, a popular means of preserving memories of deceased family members by posing them for the last time within the family group, were taken of animals and those who mourned them, and there were animal funeral services, cemeteries, gravestones, elegies, and mourning cards.[57] To some, this seemed an excessive indulgence in sentimentality, and to others, anthropomorphism, but it was also the expression of a popular, inclusive theology of animals that, in defiance of orthodoxy, granted them souls that survived death, and opened to them the gates of heaven. Paradoxically, perhaps, it was easier to mourn lives when there was some prospect of reunion, and to lay flowers on the grave of a favourite dog in anticipation of a shared life to come:

> Not hopeless, round this calm sepulchral spot,
>   A wreath presaging life we twine;
> If God be love, what sleeps below was not
>   Without a spark divine.[58]

Vivisectionists, of course, took a different view, and while they did not often discuss such matters, it seems that no-one who believed that animals had souls made a practice of vivisecting them. Perhaps some who vivisected did not believe in souls at all, for the conflict between vivisection and anti-vivisection was beginning to align itself with that of materialism versus anti-materialism (or spiritualism, if you prefer). Many of the most controversial and well publicised animal experiments involved the brain, an organ that vivisectionists treated as a mechanism, but whose subtle workings anti-vivisectionists such as Cobbe did not feel could ever be revealed by the physiologist's knife:

> The common passion for science in general and for physiology in particular, and the prevalent materialistic belief that the secrets of the Mind can be best explored in matter, undoubtedly account in no small matter for the vehemence of the new pursuit of original physiological investigations.[59]

At the turn of the century it remained a matter of great controversy whether evolution could have been responsible for the emergence of the most complex cognitive faculties, including the human capacity for love, imagination, and feeling, or whether there were some transcendent aspects of thought and consciousness that could never be explained in biological terms.[60] The two great founders of evolutionary theory disagreed: Darwin thought that evolution could account for these mental phenomena, Alfred Russel Wallace (1823–1913) that it could not.

For Christians who felt their humanity threatened by talk of the evolution of rationality, one solution was to emphasise the spiritual uniqueness of humankind, allowing that animals were rational and capable of feeling and suffering, but according humans alone the 'spark divine' of an immortal soul. It was presumably a conviction that humans were in some spiritual way ontologically different from animals that led many Christian clergy in early-twentieth century Britain to sign up for the pro-vivisection Research Defence Society (on which, see Chap. 6): they saw human lives as of intrinsically greater value to God than those of brute creation, because God had imparted something to us that mere biology could not. From a mundane perspective, most Churches had no wish to deny the theory of evolution, forcing the faithful to choose between religion and science, so they pragmatically confined the divine likeness in humankind to the immaterial part, leaving animals as mere matter, physically kin to humans, but spiritually inconsequential.

Conclusion

We have seen that, in nineteenth-century Britain, the Christian view of animals as rational but unspiritual was challenged by claims that they were either ensouled individually or were part of a collective world soul, and thus were not, as mainstream churches taught, categorically distinct from humans in a spiritual or metaphysical sense. When a few advanced followers of Pythagorean and Eastern thought proposed that the souls of animals might subsist after death, and that transmigration of souls between animals and humans might occur, this introduced a concept of the soul that was fundamentally different from that of the Christian tradition: a life-force that was constantly changing, reforming and repeating, rather than an artefact eternally linked to the human body for which it had been created.

Transcendentalism inculcated a similar, non-Christian, perspective, according to which humans were only one expression of a universal creative force of nature—arising, developing, and becoming extinct like all living things. For transcendentalists, neither humans nor animals could expect an individual afterlife, but the dynamic system in which all participated could be said to be endowed with a common soul, and death could be seen not as an end, but a return to the source that all life shared.[61] This holistic view of life on earth, the concept of mankind as transient, and the notion of a life-force common to humans and animals, were strong arguments against vivisection for those who accepted them. Though transcendentalists were sometimes condemned in the nineteenth century as atheists, their position was much closer to that of pantheism. Along with other non-Christian faiths, elements of their thinking influenced the development of the spiritual revival, and it is to this movement and its consequences for the welfare of animals in the twentieth century that we shall now turn.

Notes

1. Quoted in Rod Preece, 'Darwinism, Christianity, and the great vivisection debate', *Journal of the History of Ideas*, 64 (2003), 399–419 (Preece 2003).
2. H. Knight Horsfield, 'Old Rocket', in *The Dog in British Poetry*, ed. R.M. Leonard (London: D. Nutt, 1893), 197 (Horsfield 1893).
3. 'Have animals souls?' (editorial), *Leeds Times,* 17 May 1856, 6. (Have animals souls? 1856)
4. Preece, 'Darwinism, Christianity'; Chien-hui Li, 'An unnatural alliance? Political radicalism and the animal defence movement in late Victorian and Edwardian Britain', *EurAmerica*, 42 (2012), 1–43 (Li 2012).
5. Darwin himself took a nuanced position on vivisection, and never suggested that evolutionary theory could resolve the ethical problem.
6. Nancy Stepan, *The Idea of Race in Science: Great Britain 1800–1960* (London: Macmillan, 1982), 7 (Stepan1982).
7. Susan Merrill Squier, *Liminal Lives: Imagining the Human at the Frontiers of Biomedicine* (Durham NC: Duke University Press, 2004), 94; James Baycroft, 'Suspended animation', *Comic Annual* (London: Henry Colburn, 1842): 323–336 (Squier 2004; Baycroft 1842).
8. Li, 'An Unnatural Alliance?' (Li 2012).
9. Lansbury, *The Old Brown Dog*, ix.
10. Preece, 'Darwinism, Christianity'. For the opposing view that Darwinism 'confirmed the kinship between man and animal' see Niven, *History of the Humane Movement*, 79, and French, *Antivivisection and Medical Science.*

11. Chien-hui Li, 'An Unnatural Alliance?'; Chien-hui Li, 'A union of Christianity, humanity and philanthropy: the Christian tradition and prevention of cruelty to animals in nineteenth-century England', *Society and Animals, 8* (2000), 265–285 (Li 2000).
12. Minute book, 1832, RSPCA Archive, Southwater, CM/20, 40, 45–46, 53. (RSPCA Archive 1832)
13. Li, 'An Unnatural Alliance?' (Li 2000).
14. Drummond, *The Rights of Animals*, 15, 17; 'Society for the Prevention of Cruelty to Animals' (editorial*), Evangelical Magazine and Missionary Chronicle* 2 (1824): 357–358; Bates, 'Vivisection, virtue ethics'. (Society for the Prevention of Cruelty to Animals 1824)
15. The charitable angel's arrow was aimed, appropriately, at Parliament.
16. 'The forthcoming ass show' (editorial), *Saturday Review*, 16 (1863), 213–215. (The forthcoming ass show 1863)
17. Craig Buettinger, 'Antivivisection and the charge of zoophil-psychosis in the early twentieth century', *Historian*, 55 (1993), 277–288 (Buettinger 1993).
18. Quoted in Rod Preece, *Animal Sensibility and Inclusive Justice in the Age of Bernard Shaw* (Vancouver: UBC Press, 2011), 121 (Preece 2011).
19. Brian Harrison, 'Animals and the state in nineteenth-century England', *English Historical Review,* 88 (1973), 786–820 (Harrison 1973).
20. Li, 'An Unnatural Alliance?' It is probably no coincidence that the two English cardinals who were anti-vivisectionists, Manning and Newman, were both Anglican converts. Manning was much influenced by his wife, who died before he converted to Catholicism.
21. Robert Hull, 'On vivisection', *London Medical Gazette*, 33 (1843/1844), 219–220 (Hull 1843/1844).
22. 'Dr. Blundell's introductory physiological lecture' (editorial), *Lancet*, 9 (1825/6), 113–118; Stevenson, 'Religious elements'.
23. Revelation 6: 9–10.
24. James Lawson Drummond, 'On humanity to animals', *Edinburgh Philosophical Journal*, 12 (1831): 172–183 (Drummond 1831).
25. Anita Guerrini, 'The ethics of animal experimentation in seventeenth-century England', *Journal of the History of Ideas,* 50 (1989), 391–407 (Guerrini 1989).
26. Vyvyan, *In Pity*, 23–24.
27. Aristotle, *Metaphysics*, 1037a6.
28. Aquinas, Summa Contra Gentiles II, 68; Anthony Kenny, *Aquinas on Mind* (London: Routledge, 1993), 126 (Kenny 1993).

29. Catholic doctrine as laid down by the Council of Vienne in 1312 was, however, that '... anyone who presumes henceforth to assert defend or hold stubbornly that the rational or intellectual soul is not the form of the human body of itself and essentially, is to be considered a heretic', which appears to have excluded the possibility of a soul without a body, or a person without a soul. The heresy that the Council was trying to prevent was monopsychism: the view that all human beings shared a common intellect: Robert Pasnau, *Thomas Aquinas on Human Nature: A Philosophical Study of Summa Theologiae 1a 75–89* (Cambridge: Cambridge University Press, 2002), 159–160 (Pasnau 2002).

30. Matthew Hale, *The Primitive Organization of Mankind, considered and Examined According to the Light of Nature* (London: William Godbid for William Shrowsbery, 1677), 304, 321 (Hale 1677).

31. G.E. Bentley, 'The freaks of learning', *Colby Library Quarterly*, 18 (1982): 87–104; Douglas K. Candland, *Feral Children and Clever Animals: Reflections on Human Nature* (Oxford, Oxford University Press, 1993) (Bentley 1982; Candland 1993).

32. Daniel Tyerman, *Essays on the Wisdom of God* (London: R. Clay, 1818), 266 (Tyerman 1818).

33. J. Worthy, 'Souls and instincts of animals', *Catholic Institute Magazine* (1856): 143–149, 175–180 (Worthy 1856).

34. Richard Graves, *On the Four Last Books of the Pentateuch* (London: Henry G. Bohn, 1850), 294 (Graves 1850).

35. Preece, 'Darwinism, Christianity' (Preece 2003).

36. J. Earl Clausen (ed.), *The Dog's Book of Verse* (Boston: Small, Maynard and Co., 1916), 145–174 (Earl Clausen 1916).

37. Peter Harrison, 'Animal souls, metempsychosis, and theodicy in seventeenth-century English thought', *Journal of the History of Philosophy*, 31 (1993): 519–544; J.Z. Holwell, *Interesting Historical Events Relevant to the Provinces of Bengal, and the Empire of Indostan...* (London: T. Becket and P.A. De Hondt, 1767); see also Chi-Ming Yang, 'Gross metempsychosis and Eastern soul', in Frank Palmeri (ed.) *Humans and Other Animals in Eighteenth-Century British Culture...* (Ashgate: Aldershot, 2006), 13–30 (Harrison 1993; Holwell 1767; Yang 2006).

38. Anna Laetitia Aikin [Barbauld], 'The Mouse's Petition', *Poems* (London: Joseph Johnson, 1773), 37–40 (Barbauld 1773).

39. Peder Anker, 'A vindication of the rights of brutes', *Philosophy and Geography*, 7 (2004), 259–264 (Anker 2004).

40. Thomas Taylor, *A Vindication of the Rights of Brutes*, ed. R. Urban St Cir (Sequim, WA: Holmes, 2009), 8 (Taylor 2009).

41. House of Lords debate, *Hansard*, 15 May 1809, 14, c554.

42. Lewis Gompertz, *Fragments in Defence of Animals, and Essays on Morals, Soul, and Future State* (London: W. Hornell, 1852), 191–206; Lucien Wolf, 'Gompertz, Lewis' (rev. Ben Marsden) *Oxford Dictionary of National Biography* (Oxford: Oxford University Press, 2004), viewed 4 August 2014, http://www.oxforddnb.com/view/article/10934, (Gompertz 1852; Wolf and Lewis 2004).

43. Forster visited India, and was impressed by the animal hospitals he saw there: Thomas Forster, *Philozoia, or Moral Reflections on the Actual Condition of the Animal Kingdom* (Brussels: Deltombe and Co., 1839), viii–ix, 28, 63 (Forster 1839).

44. Thomas Forster, *An Apology for the Doctrine of Pythagoras as Compatible with that of Christianity with an Account of a New Sect of Christians* (Boulogne sur Mer: Charles Aigre 1812), 4, 7. The 'sect' was the short-lived New Christian Society of Metempsychosians (Forster 1812).

45. 'Aesculapius', The *Hospital Pupil's Guide: being Oracular Communications Addressed to Students of the Medical Profession* (London: E. Cox and Sons, 1818), 39 (Aesculapius 1818).

46. *Lancet*, 1 (1829/30), 749–753.

47. For an introduction to transcendental philosophy see Philip F. Rehbock, *The Philosophical Naturalists: Themes in Early-Nineteenth Century British Biology* (Wisconsin: University of Madison Press, 1983), 15–30 (Rehbock 1983).

48. Robert Knox, *The Races of Men: a Fragment* (London: Henry Renshaw, 1850), 426 (Knox 1850).

49. Robert Knox, *A Manual of Artistic Anatomy, for the use of Sculptors, Painters and Amateurs* (London: Henry Renshaw, 1852) (Knox 1852).

50. Bates, *Robert Knox*, 113, 132–135.

51. Ralph Waldo Emerson, *Nature and Other Essays* (Mineola, NY: Dover, 2009 [1836]), 18 (Emerson 2009).

52. Adrian Desmond, *The Politics of Evolution: Morphology, Medicine, and Reform in Radical London* (Chicago: University of Chicago Press, 1989), 183 (Desmond 1989).

53. Michael Hagner, 'The soul and the brain between anatomy and Naturphilosophie in the early-nineteenth century, *Medical History*, 36 (1992), 1–33 (Hagner 1992).

54. Bates, Robert Knox, 21, 146; Evelleen Richards, 'The "moral anatomy" of Robert Knox: the interplay between biological and social thought in Victorian scientific naturalism', *Journal of the History of Biology*, 22 (1989): 373–436; Stevenson, 'Religious elements' (Richards 1989).

55. David Mushet, *Wrongs of the Animal World* (London: Hatcher and Son, 1839), 79. Shelley thought the Ancient Greeks experienced beauty through living 'in harmony with nature': Percy Bysshe Shelley, *Essays, Letters from Abroad...* (London: Edward Moxon, 1840), 2 vols. 2, 190 (Mushet 1839);(Shelley 1840).

56. See, for example, Gerard M. Verschuuren, *Darwin's Philosophical Legacy: the Good and the Not-So-Good* (Lanham: Lexington Books, 2012), 93–100 (Verschuuren 2012).

57. Teresa Mangum, 'Animal angst: Victorians memorialize their pets', in Deborah Denenholz Morse and Martin A. Danahay (eds), *Victorian Animal Dreams: Representations of Animals in Victorian Literature and Culture* (Aldershot: Ashgate, 2007), 15–34 (Mangum 2007).

58. Francis Doyle, 'Epitaph on a Favourite Dog', in *The Return of the Guards, and Other Poems* (London: Macmillan, 1883), 264 (Doyle 1883).

59. Frances Power Cobbe, *The moral aspects of vivisection* (London: VSS, 1875), 288 (Cobbe 1875).

60. Finn and Stark, 'Medical science'.

61. Knox, *Artistic Anatomy*, 5.

References

Aesculapius. (1818). *The hospital pupil's guide: Being oracular communications addressed to students of the medical profession* (p. 39). London: E. Cox and Sons.

Anker, P. (2004). A vindication of the rights of brutes. *Philosophy and Geography, 7,* 259–264.

Barbauld, A. L. A. (1773). The mouse's petition. *Poems* (pp. 37–40). London: Joseph Johnson.

Baycroft, J. (1842). Suspended animation. *Comic annual* (pp. 323–336). London: Henry Colburn.

Bentley, G. E. (1982). The freaks of learning. *Colby Library Quarterly, 18,* 87–104.

Buettinger, C. (1993). Antivivisection and the charge of zoophil–psychosis in the early twentieth century. *Historian, 55,* 277–288.

Candland, D. K. (1993). *Feral children and clever animals: Reflections on human nature*. Oxford: Oxford University Press.

Clausen, J. E. (Ed.). (1916). *The dog's book of verse* (pp. 145–174). Boston: Small, Maynard and Co.

Cobbe, F. P. (1875). *The moral aspects of vivisection*. London: VSS.

Darwin, C. (1859). *On the origin of species by means of natural selection, or, the preservation of favoured races in the struggle for life*. London: John Murray.

Desmond, A. (1989). *The politics of evolution: Morphology, medicine, and reform in radical London*. Chicago: University of Chicago Press.

Doyle, F. (1883). Epitaph on a favourite dog. In *The return of the guards, and other poems* (p. 264). London: Macmillan.

Dr. Blundell's Introductory Physiological Lecture (Editorial). (1825/1826). Stevenson, 'Religious elements'. *Lancet, 9,* 113–118.

Drummond, J. L. (1831). On humanity to animals. *Edinburgh Philosophical Journal, 13,* 172–183.

Emerson, R. W. (2009 [1836]). *Nature and other essays* (p. 18). Mineola, NY: Dover.

Forster, T. (1812). *An apology for the Doctrine of Pythagoras as compatible with that of Christianity with an account of a new sect of Christians* (pp. 4,7) . Boulogne sur Mer: Charles Aigre.

Forster, T. (1839). *Philozoia, or moral reflections on the actual condition of the animal kingdom* (pp. viii–ix, 28, 63). Brussels: Deltombe and Co.

Gompertz, L. (1852). *Fragments in defence of animals, and essays on morals, soul, and future state*. London: W. Hornell.

Graves, R. (1850). *On the four last books of the Pentateuch* (p. 294). London: Henry G. Bohn.

Guerrini, A. (1989). The ethics of animal experimentation in seventeenth-century England. *Journal of the History of Ideas, 50,* 391–407.

Hagner, M. (1992). The soul and the brain between nature and Naturphilosophie in the early-nineteenth century. *Medical History, 36,* 1–33.

Hale, M. (1677). *The primitive organization of mankind, considered and examined according to the light of nature* (pp. 304, 321). London: William Godbid for William Shrowsbery.

Harrison, B. (1973). Animals and the state in nineteenth-century England. *English Historical Review, 88,* 786–820.

Harrison, P. (1993). Animal souls, metempsychosis, and theodicy in seventeenth-century English thought. *Journal of the History of Philosophy, 31,* 519–544.

Have Animals Souls? (Editorial). (1856, May 17). *Leeds Times,* 6.

Holwell, J. Z. (1767). *Interesting historical events relevant to the provinces of Bengal, and the Empire of Indostan...* London: T. Becket and P.A. De Hondt.

Horsfield, H. K. (1893). Old rocket. In R. M. Leonard (Ed.). *The dog in British poetry* (p. 197). London: D. Nutt.

Hull, R. (1843/1844). On vivisection. *London Medical Gazette, 33,* 219–220.

Kenny, A. (1993). Aquinas, summa contra Gentiles II, 68. *Aquinas on mind* (p. 126). London: Routledge.

Knox, R. (1850). *The races of men: A fragment* (p. 426). London: Henry Renshaw.

Knox, R. (1852). *A manual of artistic anatomy, for the use of sculptors, painters and amateurs.* London: Henry Renshaw.

Li, C. (2000). A union of Christianity, humanity and philanthropy: The Christian tradition and prevention of cruelty to animals in nineteenth-century England. *Society and Animals, 8,* 265–285.

Li, C. (2012). An unnatural alliance? Political radicalism and the animal defence movement in late Victorian and Edwardian Britain. *EurAmerica, 42,* 1–43.

Mangum, T. (2007). Animal angst: Victorians memorialize their pets. In D. D. Morse & M. A. Danahay (Eds.), *Victorian animal dreams: Representations of animals in Victorian literature and culture* (pp. 15–34). Aldershot: Ashgate.

Mushet, D. (1839). *The wrongs of the animal world.* London: Hatchard and Son.

Pasnau, R. (2002). *Thomas aquinas on human nature: A philosophical study of summa theologiae 1a 75–89* (pp. 159–160). Cambridge: Cambridge University Press.

Percy Bysshe Shelley. (1840). *Essays, Letters from Abroad...* (Vol. 2, p. 190). London: Edward Moxon.

Preece, R. (2003). Darwinism, Christianity, and the great vivisection debate. *Journal of the History of Ideas, 64,* 399–419.

Preece .R (2011). *Animal sensibility and Inclusive justice in the age of Bernard Shaw* (p. 121). Vancouver: UBC Press.

Rehbock, P. F. (1983). *The philosophical naturalists: Themes in early-nineteenth century British biology.* Wisconsin: University of Madison Press.

Richards, E. (1989). The "moral anatomy" of Robert Knox: The interplay between biological and social thought in Victorian scientific naturalism. *Journal of the History of Biology, 22,* 373–436.

RSPCA Archive. (1832). *Minute book* (CM/20, pp. 40, 45–46, 53). Southwater.

Society for the Prevention of Cruelty to Animals (Editorial). (1824). *Evangelical Magazine and Missionary Chronicle, 2,* 357–358.

Squier, S. M. (2004). *Liminal lives: Imagining the human at the frontiers of biomedicine.* Durham, NC: Duke University Press.

Stepan, N. (1982). *The idea of race in science: Great Britain 1800–1960* (p. 7). London: Macmillan.

Taylor, T. (2009) *A vindication of the rights of brutes.* R. Urban St Cir (Ed.). Sequim, WA: Holmes.

The Forthcoming Ass Show (Editorial). (1863). *Saturday Review, 16,* 213–215.

Tyerman, D. (1818). *Essays on the wisdom of God* (p. 266). London: R. Clay.

Verschuuren, G. M. (2012). *Darwin's philosophical legacy: The good and the not-so-good* (pp. 93–100). Lanham: Lexington Books.

Wolf, L. (2004). *Oxford dictionary of National Biography* (G. Lewis, rev. M. Ben). Oxford: Oxford University Press. Retrieved August 4, 2014, from http://www.oxforddnb.com/view/article/10934.

Worthy, J. (1856). Souls and instincts of animals. In *Catholic Institute Magazine*   (pp. 143–149, 175–180).

Yang, C.-M. (2006). Gross metempsychosis and Eastern soul. In F. Palmeri (Ed.), *Humans and other animals in eighteenth-century British culture...* (pp. 13–30). Ashgate: Aldershot.

4

A New Age for a New Century: Anti-Vivisection, Vegetarianism, and the Order of the Golden Age

To Josiah Oldfield (1863–1953) belongs the distinction of having founded Britain's, and quite possibly the world's, first anti-vivisection hospital, the short-lived Hospital of St Francis, which opened in 1898 at 145 New Kent Road in South London. Oldfield is now remembered, if at all, as a pioneering dietary reformer; a bearded, Bible-quoting, besmocked prophet of fruitarianism who devoted his considerable intellectual energy to a raft of utopian projects: a vegetarian hospital, a fruitarian colony, a programme of dietetics. To his critics he was a crank, but it was not easy to dismiss the arguments of a Middle Temple barrister, medical graduate and Oxford Doctor of Law. Oldfield's medical career was an unusual one, pursued outside conventional hospital circles and devoted to a health reform programme whose principles included a cruelty-free diet, a more natural lifestyle, and an emphasis on spiritual as well as physical health. His work is considered here for the light it sheds on a broader health and spiritual reform movement that drew on influences as diverse as Eastern philosophy, transcendentalism and Darwinism to promote a worldview of 'universal kinship' and harmony with nature.[1] This movement, which included vegetarians, anti-vivisectionists and other social improvers, approached the new century in anticipation of a new age in which materialism would be tempered

© The Author(s) 2017
A.W.H. Bates, *Anti-Vivisection and the Profession of Medicine in Britain*,
The Palgrave Macmillan Animal Ethics Series, DOI 10.1057/978-1-137-55697-4_4

by spiritualism and true science would flourish, heralding an age of prosperity in which violence and oppression, including the abuse of animals, would no longer have a place.

Oldfield set about building this utopia by establishing 'healthian' colonies in the South East of England—the warmest region and thus the best suited to the outdoor work, fruitarian diet, nudity and sun baths that his followers were encouraged to enjoy. The locals regarded these communes of free-living fruitarians with suspicion, but Oldfield's strategy also included vegetarian and anti-vivisection hospitals that he hoped would demonstrate the practical benefits a meat- and cruelty-free lifestyle. His projects were typical of the new age movement in that, while attractive to a minority, they failed to win enough support among people of influence or achieve the necessary level of popular acceptance to bring about significant social change. The first world war undermined public confidence in the prospect of creating a cruelty-free utopia, but the final straw for the back-to-nature movement was the subsequent co-option of 'green' ideology by British fascists, which, along with its links with German National Socialism, made its values seem subversive and treasonable as Britain once more prepared for war.

The Food Reform Movement in Britain

The antecedents of the back-to-nature movement lay in social vegetarianism, of which Oldfield was a lifelong champion. In Britain, organized vegetarianism had been linked from the outset with Fabianism and, in particular, the Concordium (1838–1848), a utopian socialist community that collectively sought the inspiration of the 'Triune Universal Spirit', and whose journal, the *New Age, Concordium Gazette, and Temperance Advocate*, placed health at the centre of an idealistic programme of anti-militarism, temperance and 'vegetarianism' (veganism in modern terminology), which was extoled as the 'beginning and end of all true reforms'.[2] The Concordium, in turn, traced its roots to the American transcendentalist movement—which had a much broader social emphasis than the scientific transcendentalism taught in British medical schools—and named its Surrey headquarters after the New England transcendentalist philosopher Amos Bronson Alcott (1799–1888).

The ethical socialists of the Concordium were concerned about animal welfare for different reasons than the anti-cruelty societies. While the latter were worried that working class cruelty, left unchecked, would spread and threaten the stability of society, socialists thought that cruelty was imposed from above, and that the harsh dominance of mankind over animals both mirrored and encouraged the exploitation of the poor by the rich. Though the Concordium lasted only 10 years, after its demise other socialist groups took up the cause. One of these was the Humanitarian League, founded in 1891 by Henry Stephens Salt (1851–1939), who became its General Secretary and editor of its journals. The League opposed the infliction of avoidable suffering on any sentient being, campaigning against corporal and capital punishment and blood sports as well as vivisection. Salt himself was an ethical vegetarian, an anti-vivisectionist and a pacifist—a not uncommon combination among socialists—who believed that the new-found kinship with animals that had been revealed by Darwin's theory of evolution warranted the extension of rights to the non-human 'races'. Animal rights was something of a surrogate cause among socialists, because they assumed, with reasoning the reverse of Thomas Taylor's, that if animals were recognised as having rights, humans could not possibly be denied them:

[The] notion of the life of an animal having 'no moral purpose', belongs to a class of ideas which cannot possibly be accepted by the advanced humanitarian thought of the present day—it is a purely arbitrary assumption, at variance with our best instincts, at variance with our best science, and absolutely fatal (if the subject be clearly thought out) to any full realization of animals' rights. If we are ever going to do justice to the lower races [*i.e.*, animals], we must get rid of the antiquated notion of a 'great gulf' fixed between them and mankind, and must recognize the common bond of humanity that unites all living beings in one universal brotherhood.[3]

Perhaps the most prominent of the organizations that carried on the Concordium's work after its closure were the Vegetarian Society and the Order of the Golden Age, both of which aimed to improve health and morals by introducing a lifestyle that was less cruel and more in harmony with the natural world, and to both of which Oldfield would make a significant contribution.

Oldfield joined the Vegetarian Society in the 1860s while he was a theological student at Oxford. The Society had been founded in 1847, primarily to promote vegetarianism on health grounds, but from the beginning it had political and religious overtones. Its membership, which was never more than a few hundred, included ex-members of the Concordium, doctrinally vegetarian Christians such as Cowherdites, and a handful of undergraduates prepared to sign up to a somewhat controversial cause. The Oxford branch served as a kind of club for left-leaning social crusaders, and joining it was a particularly provocative move for a theological student, since ethical vegetarianism seemed to run counter to the conventional Christian wisdom that animals had been placed on earth for the benefit of mankind.[4]

Oldfield's motivation, however, seems to have lain primarily in social concerns: he hoped that vegetarianism would combat poverty and ill health (by encouraging the poor to spend more on vegetables and less on strong drink[5]), and that putting a stop to the cruel slaughter of animals would lead to a more peaceful and humane society. He carried his desire for social improvement into practice after finishing his theological studies, declining to take holy orders as expected and instead pursuing a career first in law and then in medicine, but he also did his best to relieve the sufferings of animals, publishing *A Groaning Creation*, a characteristic blend of logical argument and impassioned rhetoric, followed by *A Tale of Shame and Cruelty*, in which he described the torments to which animals were subjected as they were transported and slaughtered for food, and advocated vegetarianism, 'a natural and humane diet', as the way to avoid it.[6]

Towards the end of the nineteenth century, Oldfield also joined the Order of the Golden Age, a Christian dietary reform movement founded in 1882 with four grades of membership: the lowest required members to believe in the Apostles' Creed, rise early, dress soberly, and be 'humane', while the higher grades required progressive abstinence from meat, fish and alcohol.[7] The Order organized vegetarian banquets and lectures 'for the furtherance of our propaganda',[8] its stated aims being '[t]o proclaim a message of Peace and Happiness, Health and Purity, Life and Power', and '[t]o hasten the coming of the Golden Age when Love and Righteousness shall reign upon earth... by proclaiming obedience to the laws of God'.[9]

Membership was more controversial than it might appear, to the extent that the family of one early member suspected (wrongly, as it turned out) that her link with the Order had been the cause of her being committed to a lunatic asylum.[10] To the uninitiated, vegetarianism seemed an irrational practice, and the Order's mission to live in peace and harmony with the animal kingdom, ridiculous: one newspaper suggested that mad dogs, runaway bulls and tigers should be allowed to join.[11] Those people who did enlist saw themselves as misunderstood pioneers, even revolutionaries, and hoped eventually to convert millions to the 'simpler habits of life', and so transform society by ending food shortages and ushering in a 'Reign of Plenty' that would put an end to war and disease.[12]

As a lawyer, Oldfield accepted that animals had some rights, such as 'the inherent right of the non-human races to be exempted from the infliction of pain...',[13] and he campaigned actively for human rights, founding the Society for the Abolition of Capital Punishment, touring India to study the workings of its legal system, and publishing *Hanging for Murder* (1908) in a bid to get the law on judicial execution changed. It is significant, therefore, that he never called for legislation to protect laboratory animals, presumably because he expected to end vivisection not in the courts, but by bringing about a cultural change that would restore humankind to a more natural and compassionate relationship with the animal world—a cause to which he would devote his life.

The Oriolet Vegetarian Hospital

Having abandoned law for medicine, and while still a medical student, Oldfield set himself up as 'warden' of a pioneering vegetarian hospital that proved popular with patients, though it did not escape the controversy that consistently dogged his endeavours. The Oriolet Hospital, a converted villa with spacious gardens in Loughton on the Eastern outskirts of London, opened in 1895 with an endorsement from the Order of the Golden Age.[14] With the assistance of a visiting medical officer, Oldfield admitted a total of 190 patients in its first year: men, women and children suffering from everything from eczema to varicose veins to paraplegia, all of whom received 'dietetic treatment'. So optimistic was Oldfield that he could also cure carcinoma, sarcoma and epithelioma with a vegan diet that he advertised for patients with these conditions, who were to be admitted for free.[15]

The meat- and alcohol-free diet was apparently well tolerated, as it was never found necessary to vary the rule that no 'fish, flesh, or fowl' was served. What started as the complaints book was soon filled with compliments from patients who showed a suspiciously good grasp of the hospital's purpose: one saw it 'as a proof of what the Vegetarian diet and Hygienic principles properly carried out, will do for suffering humanity', while another hoped that vegetarianism would become 'widely known and recommended... as I am sure it will be by all who have given it a fair trial'. Oldfield's local appeals for his 'pioneer hospital in humane dietetics' attracted gifts of everything from fruit and vegetables to framed Bible verses, but the hospital relied for financial support primarily on its chairman, the shipbuilder Arnold Frank Hills (1857–1927), himself an ardent vegetarian and teetotaller, who kept the it afloat by contributing hundreds of pounds a year.[16]

The medical profession of the time generally disapproved of dietary therapy, and though Oldfield graduated LRCP, MRCS in 1897 and was duly entered in the medical register, he was not welcomed into the fold. According to a critical piece in the BMJ, aimed at the Oriolet, hospitals that relied on 'some special fad or other as to diet...' tended to do well only because they attracted likeminded patients who had faith in the treatments they received there, but they were actually a kind of 'medical sack racing', because patients got better in spite of the restrictions rather than because of them. The British Medical Association (BMA), which published the BMJ, was essentially a trade union, whose defence of its members' interests included opposing 'faddism' wherever they found it: 'Abstinence from animal food is one of these fads, abstinence from alcohol is another. We have not yet heard of a hospital founded on the principle of abstaining from the use of opium...'. They admitted that the results from the Oriolet seemed 'perfectly good', but concluded that '[a]ll these one-legged institutions are tarred with the same brush in this respect, that the patient in choosing his hospital chooses his treatment, which is ethically wrong'.[17]

The same might have been said of any hospital, but vegetarianism troubled the BMA because they thought doctors were recommending it on religious or socio-political grounds rather than medical ones. Whether a regime worked in practice was immaterial if it was chosen for the wrong reasons: what the BMA was opposed to was ideological medicine, insisting that the individual patient's best interests must be addressed disinterestedly in every case. Of course, one could find many examples where mainstream medicine was as ideological as any of the alternatives, but the principle that doctors should not impose their own moral values on patients left vegetarians and anti-vivisectionists vulnerable to accusations that they were pushing their own ethical agenda. The Master of the Rolls (Chief Justice of the Court of Appeal) did, however, dismiss an objection from the profession's leaders that a vegetarian hospital should not operate as a charity because its primary purpose was 'the propagation of a fad' rather than the treatment of the sick, though the courts later came to the opposite conclusion with regard to anti-vivisection charities, with dire consequences for their funding.[18]

The Hospital of St Francis

In March 1897, Oldfield announced a plan to open an anti-vivisection hospital:

> In commemoration of the Queen's Jubilee, the anti-vivisectionists of this country and the Continent have decided to found a hospital on what they call purely humanitarian lines. It is meant to be a protest against 'all forms of cruelty and especially of vivisection'. It is proposed to call it 'The Hospital of St. Francis', in memory of Saint Francis of Assissi [sic]. It will be built in the south of London, where the need for a general hospital is very pressing.[19]

Although he was, at the time, a member of the Executive Committee of the Victoria Street Society, and despite the lofty allusion to international anti-vivisectionists, St Francis's was Oldfield's personal initiative. As such, the plan was characteristically both idealistic and shrewd:

a South London Hospital would fly the flag for anti-vivisection at the same time as helping the poor who, in a district desperately short of hospital beds, would gladly accept any treatment offered to them. In such a deprived area, suitable (or, as it turned out, unsuitable) premises could be acquired relatively cheaply, and as St Francis's was the only anti-vivisection hospital in Britain, there would be no alternative for donors who wished to support a hospital while being certain they were not funding vivisection.

Oldfield was a persuasive and determined fundraiser who cast his nets widely. Public feeling against 'vivisecting hospitals' was running high after a pamphlet campaign against them by the leading anti-vivisectionist Stephen Coleridge (1854–1936), and Oldfield appealed to potential donors' religious fervour, exhorting them to '… rise in your millions and pour into the crucible of healing your golden rings…', in order to '… build a fair and beauteous temple of healing'.[20] This was a time when devout ladies were known to give up their jewels to adorn the sacred vessels in Anglo-Catholic churches, and St Francis's, as its name proclaimed, was manifestly a Christian institution, 'an aspiration after the gentleness of the divine' that could 'brook no delay' because those who supported it were 'On the King's [i.e., Jesus Christ's] business'.

Although Oldfield's own faith was idiosyncratic and barely containable even within the very broad limits of Anglicanism, it had every appearance of being heartfelt. He spoke, wrote and even looked— white-bearded, white-suited, and white-coated—like a prophet charged with bringing God's message of compassion to the world. 'No man having the Christ-Spirit within his heart, can see animals ill-treated without a protest!',[21] he thundered, but though he made every oratorical effort to persuade local people to support the hospital out of generosity of spirit, he shrewdly threw in an appeal to self-interest, promising humane treatment, in contrast to the abuses and indifference he cleverly implied awaited any patients of limited means who found themselves in a teaching hospital bed at the mercy of experimentalists:

> Let no demand for 'material' [as teaching hospitals sometimes tactlessly described their patients] ever sully the beauty of [the Hospital's] teaching. Let every patient be looked upon as a casket of priceless worth…. Let no

shadow of inflicted pain upon compulsory victims shut out the sunlight of God's grace... the dying hours shall be sacred, and the body, though the gentle spirit has passed on, shall still be a thing of reverence.

For potential donors, the alternative was stark: money given to the Prince of Wales's Fund (later the King's Fund), the principal charity that distributed money to London's voluntary hospitals, would, according to Oldfield, 'go to strengthen the state that exists and to perpetuate things as they are'.[22]

This did not endear him to the charity fraternity, and as early as 1898, before his hospital had even opened, the Charity Organization Society (COS), a semi-official watchdog, stepped into investigate his fundraising efforts. Their inspector, Charles Carthew, was unimpressed by Oldfield's London office, where his representative, 'a young man got up á la Bohemian', seemed to know little about the proposed hospital, even mixing up anti-vivisection with anti-vaccination (the two were not yet linked, though they would become so).[23] When Carthew finally met Oldfield he came away with the impression, as many others did, that he was 'not altogether straight', though his investigations discovered only that Oldfield was a barrister with chambers in Mitre Court, resident medical officer to the Oriolet Hospital in Loughton, and the author of monographs on 'Tuberculosis' and 'Starch as a food in nature'.[24] The COS concluded he was acting in 'good faith', but advised donors not to support his hospital on the grounds that it was likely to prove of scant public benefit.[25] More candidly, they told Walter Vaughan Morgan (1831–1916), a potential donor who would later become a committee member of the National Anti-Vivisection Hospital (as well as Lord Mayor of London and a baronet), that St Francis's was being set up for Oldfield's 'private purposes', and that he had a bank account jointly in his own name and that of the hospital. This disclosure cost Oldfield Vaughan Morgan's support, and the COS presumably gave similarly discouraging replies to other enquirers.[26]

That the hospital would be of little public benefit was true in so far as it had too few beds to make an appreciable difference to the sick of South London, but Oldfield intended it primarily as propaganda for the anti-vivisection cause, and even a small but flourishing hospital would have sufficed to show that cruelty-free medicine was a viable prospect. Regrettably, the official opening of the hospital, in April 1898, went almost unnoticed, and when the COS inspector arrived unannounced shortly afterwards, he found there were no patients.[27] Oldfield was still busy seeking sponsors, having already persuaded Stephen Coleridge to become the hospital's chairman, and having recruited an impressive number of vice-presidents, including the Duke of Beaufort, Lords Llangattock (1837–1912) and Harberton, and the Dowager Countess of Portsmouth (1834–1906).

The number and quality of the hospital's patrons, most of whom probably never even visited it, was conspicuously disproportionate to its facilities: after it had been open for two years there were still only eleven beds, amply served by three medical officers in addition to Oldfield, and overseen by a matron.[28] The medical officers were obliged to forswear vivisection but the converted town house in which the hospital was located—an unimposing, narrow, redbrick building next door to a bicycle factory—possessed no laboratories; the purpose of the pledge was to demonstrate their humane principles.[29] There was barely enough money to keep the tiny anti-vivisection hospital open, but it could claim the distinction of being Britain's first.

Unfortunately, the haste with which St Francis's was set up probably did its cause more harm than good. The wards were cramped and shabby, a failing not lost on its critics, foremost among whom was the financier and doyen of the voluntary hospital system, Sir Henry Burdett (1847–1920), who published a damning report in his journal, *The Hospital*, which described St Francis's as a 'wretched, grubby little house' with fittings of a 'poverty-stricken character'. It was obvious to him that this 'curious excrescence on London charity' was run 'not for the benefit of the patients' but so that 'the possibility of treating disease on a non-meaty diet might be demonstrated'.[30] Burdett seems to have confused it with the Oriolet, an understandable mistake since Oldfield was best known as a vegetarian, a rare thing for a doctor at the time, though the staff of St Francis's were, as Burdett himself noted, permitted to order meat if their patients wanted it. It is difficult to dissent from Burdett's judgement that: '... the prospects of the institution have been sacrificed to the ambition of those in power on its council to be able to say that there is at least one hospital in London which definitely excludes vivisection', but it is telling that, although only dedicated anti-vivisectionists would have been likely to give to St Francis's, Burdett, London's greatest hospital fundraiser, was concerned about the precedent that might be set by diverting even this tiny amount of charity from the many hospitals that came under his financial control.

Burdett's attack on a hospital with fewer than a dozen beds was as disproportionate as the support the hospital attracted: its ten vice-presidents and thirteen patronesses lent their aristocratic names rather than their money, but their social cachet helped draw attention to Oldfield's project. If donations from philanthropists opposed to vivisection could fully support this one, independent, cruelty-free hospital, then there might be scope for more, and the voluntary hospitals would begin to find themselves poorer. The signs were worrisome for the orthodox: the vicar of St John's church in Westminster, a supporter of the anti-vivisection cause, held his usual collection for the Hospital Sunday Fund, which went to London's voluntary hospitals, followed by a separate collection just for the Hospital of St Francis, which raised over five times as much.[31] The OGA's periodical, *The Herald of the Golden Age*, probably the most widely circulated vegetarian magazine in the English-speaking world, urged its readers to 'let your church collection plate pass by if you are doubtful whether they are sound on vivisection'.[32] It did not matter that most hospitals that received money from the Sunday Fund did not experiment on animals and had no facilities to do so, in the eyes of the scrupulous they were all tarred with the same brush.

Oldfield sent out begging letters to everyone from Dukes to Aldermen, some of whom passed them on to the COS, whose inspector concluded that 'On the whole I do not think there is anything very definite that can be said against Dr Oldfield.... He is a qualified Doctor and the mere fact that he is a rabid vegetarian is not in itself to his discredit', thought the COS did its best to discourage donors, advising them to '...leave Dr Oldfield and all his works entirely alone'. Between them, Burdett and the COS succeeded in stifling Oldfield's struggling venture; in 1904 he announced a last-ditch plan to relocate the hospital to Camberwell Green, but the £2000 needed for the move was not forthcoming and it closed, the remaining funds being transferred to the newly opened National Anti-Vivisection Hospital in Battersea.[33] According to a press report, St Francis's had treated over 100,000 out-patients and 428 in-patients in just under six years.[34]

Back to Nature

Oldfield next turned his attentions to creating a hospital in the country where patients could receive dietary treatment in healthy natural sur-roundings. The Lady Margaret Hospital in Kent offered fresh air and 'dainty fruitarian meals', but the more esoteric aspects of its programme began to arouse suspicions.[35] Oldfield's links with the Order of the Golden Age were well known, and an anonymous correspondent, per-haps confusing it with the esoteric Order of the Golden Dawn, told the COS that Oldfield had connections with Swami Laura and Theodore Horos, a husband and wife team of serial fraudsters whose Theocratic Unity Temple had been the subject of a financial and sexual scandal two years earlier.[36] In search of evidence, the COS approached the Medical Defence Union, a mutual insurance society for medical practitioners, of which Oldfield does not seem to have been a member. Nevertheless, their representative, Dr Bateman, had apparently heard of him by repu-tation and was more than happy to pass on a torrent of gossip: Oldfield was married but his wife refused to live with him because he was a 'crank' and a 'sexual pervert [in this context, a womaniser]', and in Kent he had 'got hold of a lot of silly, foolish women and could do just what he liked with them'. For good measure, Bateman told the COS: 'You can't trust a fellow who lives on nuts ... it only makes them more and more earthly'.[37]

By this time, Oldfield had antagonised the medical profession not only by using a meat-free diet as therapy and campaigning against vivisection, but also by flouting professional and social standards. The *Lancet* complained that he arranged for favourable news stories about himself to appear in popular newspapers such as the *Daily Mirror* and *Penny Magazine* in order to publicise his hospitals, which was almost, but not quite, the cardinal medical sin of self-advertising.[38] It appears, however, that Oldfield, as one might expect of a barrister, was adept at sailing close to the wind. That he was never reported to the General Medical Council (GMC), still less investigated by them, despite his unpopularity with some sections of the profession and their paymasters, who were presumably watching hawk-like for him to slip up, surely indicates that there was no substance to the rumours of misconduct. As the GMC would not have ignored complaints from Oldfield's patients, it can be confidently stated that they made none.

Lady Margaret's was scarcely a hospital in the medical sense at all, since its regime relied mostly on healthy living rather than therapeutics. Though it retained a link to the more conventional Margaret dispensary in London, by 1908, the 400-acre site was known as Margaret Lodge Colony, and its proprietor not as 'Dr Oldfield' but 'Mr Warden'. A representative from the COS found 'bareness, cleanliness & want of comfort'. Though there was a farm that kept residents supplied with fresh milk, butter and eggs, the spartan, meatless regime came as a surprise to some new residents: one described the communal accommodation as a 'cowshed' (it was actually a former oast house), and another found the food 'very nasty'.[39] Children brought from the London slums to spend a summer helping on the farm left with their health apparently improved by clean air and fresh food, though it did not always appear so, since after roaming freely in the fields and woods for months they arrived home more ragged than ever.[40]

Heralding the Golden Age

Oldfield's work in promoting vegetarianism, anti-vivisection, and health reform was all part of his commitment to bringing in a golden age, to which all these other causes contributed. Vegetarianism and anti-vivisection in particular were close allies: when, in 1880, the anti-vivisection campaigner Anna Kingsford (1846–1888) submitted her thesis for a Paris medical degree (which she had scrupulously completed without recourse to vivisection), she chose to write it on vegetarianism, a less inflammatory choice than anti-vivisection, though still sufficiently controversial for her to be refused the customary public defence of her work. As Vyvyan observed, Kingsford probably intended her anti-vivisection message to be read between the lines of her argument for vegetarianism, since the objections to vivisection and meat-eating (cruel, brutalising, spiritually coarsening) were essentially the same.[41] When it came to fundraising, anti-vivisectionists and vegetarians often worked together: the committee appointed by the VSS to raise funds for an anti-vivisection hospital included the President of the Vegetarian Society, Ernest Bell (1851–1933).

According to one disgruntled Medical Officer of Health, writing in 1902, there was a distinctive personality type, which he called 'the anti', that was common to, among others, anti-vivisectionists, vegetarians, teetotallers and advocates of artificial contraception: '[he] is frequently a nonconformist in religion, usually a supporter of the Opposition in politics, and his chief recreations are crusading and the smashing of idols'.[42] The 'anti' was not confined to a particular social group—he, or she, was as likely to be found among the aristocracy as the working classes—but most were radicals in the true sense of the word, that is to say, they believed that human priorities needed to be re-evaluated and reformed from the ground up. The 'antis' included socialists, feminists, pacifists, and others disenchanted with a culture of industrialisation, urbanisation, and capitalism, whose calls for a return to a more natural way of living—the inspiration for the twentieth century back-to-nature movement—included a boycott of vivisection and meat-eating, not merely because these things were harmful to animals, but because a society preoccupied with the flesh—whether consuming it for food or vivisecting it in search of answers—was thought unlikely to grow spiritually, which the reformers thought an essential prerequisite for the desired social transformation.

It was to this end that the Order of the Golden Age was 'reconstituted' in 1904 under the presidency of Sidney Hartnoll Beard (1862–1938), with Oldfield on its six-strong General Council.[43] Beard saw the fight for more humane treatment of animals both as part of the new age programme and his Christian duty: 'the supremacy of Love and Gentleness, Spirituality and Mercy' proclaimed by Jesus ought to be extended, he argued, to 'sub-human' creatures, who were to be treated with 'beneficence'.[44] Oldfield agreed, writing that Christians should eschew all forms of killing, including butchering animals to celebrate

Christmas.[45] Ending the ill-treatment of animals in the farm, the slaughterhouse, and the laboratory was a precondition for realizing the Golden Age because cruelty, killing, and carnivorism were inherently unspiritual:

> Is it any wonder that our spirituality is at such a low ebb; that we are floundering in a slough of materialistic agnosticism and nescience; that we are in bondage to disease and the fear of death; that the barrier which separates us from the spiritual world is an opaque wall rather than a transparent veil; that the angels and ministering spirits of the higher spheres, either cannot, or will not, commune with such a carnal race of beings; that genuine spiritual experience and conscious realization of the Divine Presence and Influence, are so rare amongst us that such things are scarcely ever mentioned in our Churches ...[46]

The *Herald of the Golden Age* tried to mobilise opposition to medical vivisection, publishing a condemnation by the surgeon Robert Howell Perks (1855–1929), who wrote that it should be 'regarded as a criminal offense upon Earth—as it already is in Heaven', and an editorial which said that the reported 'indifference' and 'laughter' of students at University College was proof that vivisection demonstrations led to 'hardening of heart and searing of sensitive feeling'.[47] Its suggestion for stemming animal experiments was 'closer inquisition into the [disposition of the] hospital funds'.[48]

The OGA's opposition to vivisection alone would have been enough to earn it the disapprobation of the medical profession, had they not already been hostile to its vegetarianism. To members of the OGA, it was necessary, in order to reach the higher spiritual levels, to abjure the flesh-eating habits of wild animals and primitive men.[49] The orthodox medical view, however, was that meat eating was essential to sustain physical health, and that vegetarianism was a dangerous trend. In 1853, the *Lancet* had reported the 'recovery' of a vegetarian opium-eater ('a little, withered creature') after the restoration of an animal diet.[50] Over 50 years later, that journal still considered vegetarianism incompatible with vigorous health, suggesting in an editorial that its prevalence among 'oriental' peoples, a point often positively adverted to by its supporters, might explain 'the marked superiority of the European', and the fact that 'men have often to be employed in India for work that women will do in England…'.[51]

Vegetarianism was condemned as un-British, un-Christian, and disloyal to one's fellow humans, for placing their interests and those of animals on almost the same level. The OGA declared itself 'above all things a society of Christians', but rather than claiming the traditional 'dominion' over animals, took its inspiration from the Old Testament prophecy of a 'Messianic Age' (the 'peaceable kingdom') in which all creatures would live in harmony and killing for food would cease. The Golden Age would be achieved when this perfect state of living, the desire for which remained latent in the human psyche,[52] was finally restored:

> The wolf also shall dwell with the lamb, and the leopard shall lie down with the kid; and the calf and the young lion and the fatling together; and a little child shall lead them. And the cow and the bear shall feed; their young ones shall lie down together: and the lion shall eat straw like the ox. And the sucking child shall play on the hole of the asp, and the weaned child shall put his hand on the cockatrice' den. They shall not hurt nor destroy in all my holy mountain: for the earth shall be full of the knowledge of the Lord, as the waters cover the sea.[53]

This divinely-mandated state could only be relaized after an extensive programme of social reform had swept away vivisection, meat-eating and human conflict. At present, the labourers were few, but the work was God's will; as Oldfield wrote in a flyer for his anti-vivisection hospital: 'The whole creation is groaning and travailing in anguish, and praying to be delivered from the body of death…. Now is the epoch moment to stamp the coming century for Humanity'.[54]

Harmony with Nature

In common with the wider new age movement of the late-nineteenth and early-twentieth century, the OGA's goal was to restore the pre-eminence of the spiritual in all aspects of life. With regard to science, this meant not protesting against it but working towards a closer union between scientific and spiritual thought. Medical science in particular, they felt, had become too wedded to materialism; its practitioners might follow a personal religion, but their faith and their experiments occupied different worlds; they were, to borrow a modern phrase, 'non-overlapping magisteria'. There was a crucial distinction to be made between 'so-called' science, practised mostly in the laboratory and constrained within narrow parameters, and 'true science', which understood the world holistically by combining observation and experiment, faith and feeling.[55]

Darwin's theory of species change, for example, had helped many people to understand what transcendentalists and others claimed to have known intuitively: that 'all life is *one*'. This principle was central to the OGA's mission, and one way for its Christian membership to affirm it without compromising their status as adoptive children of God was to reconceptualize non-human animals as 'living souls' with their own hopes, joys and sorrows, 'similar to our own', and a similar capacity for virtue: according to one contributor to the *Herald*, a dog that licked the hand of a vivisector was as good a moral exemplar as any of the 'imaginary saints'.[56]

The reformers' call for faith and sentiment to guide science in all its aspects, including medicine, was of course ignored by most experimentalists, who preferred to keep external interference to a minimum. According to Oldfield, however, medical scientists already allowed their beliefs to influence their work, though without admitting it: 'it is absolutely unscientific', he wrote, 'to talk about the necessity of sacrificing a thousand dogs or guinea-pigs if need be to save one human life, because we do not know the comparative values about which we are pretending to dogmatize'. Vivisectionists, he claimed, assumed a priori that animals' lives had a lower value than human ones, but had no scientific justification for their position; it could be challenged, and Oldfield did so: 'I have seen a semi-human dog and I have seen a semi-reptilic imbecile man, and … I should have estimated the life of that so-called dog to be of more value than the life of that so-called man'. A few pages later, he spelled it out even more bluntly: 'Some non-humans may be of more value than some humans…'. His characteristically immodest but ingenious argument was that, as the value of animals' lives could only be judged intuitively and not scientifically, those best able to do so were those, like himself, who had attained a 'higher' awareness of nature:

> The higher science … is always reverent in the presence of the mystery of life…. The higher the man the more nearly he approaches to those heights of scientia and gnosis, which are the crowning stamp of the true scientist, the more reverence he has for his fellow traveller – a true brother in the eyes of science – on the same spiral pathway of vitality, towards a perfection of evolution.[57]

The key question was how to acquire this profound understanding: the Golden Age was gnostic in the sense that the deeper knowledge that Oldfield and others laid claim to could not be grasped by all, at least not in the current state of the world, but was achievable only after long study and reflexion, and with the benefit of spiritual insight. One way to obtain the latter was to study other spiritual and religious traditions, and in practice the OGA's theology tended towards syncretism, with some of its members anticipating that the future would see the establishment of a 'world religion'.[58] These theological developments had possible benefits for animals: it was suggested that the failure of Christians to recognise that animals had souls that survived death placed them 'on a lower spiritual plane' than Buddhists, and that by making a leap of faith and accepting the possibility of human and animal reincarnation, Christians could begin to strive for a better life for all creatures on Earth, rather than selfishly working towards their own salvation.[59]

The influence of Eastern philosophy and religion was also mediated through theosophy, which was closely linked with anti-vivisection. The Theosophical Society and the Victoria Street Society were founded in the same year, and had common purposes and supporters to the extent that, according to Vyvyan, they were practically sister movements. The anti-vivisection and vegetarian doctor Anna Kingsford was instrumental in converting the prominent theosophist Annie Besant to the humane movement, and in turn was herself converted to theosophy, becoming president of the Theosophical Society's London branch in 1883, and launching a psychic war against the vivisectionists Paul Bert (1833–86), Claude Bernard and Louis Pasteur, a campaign in which she claimed some success.[60]

Many other prominent vegetarians were active theosophists, including the Vegetarian Society's London secretary (and ex Concordium member), George Dornbusch (1819–1873); Constance Wachtmeister (1838–1910, a close friend of Blavatsky), and the homoeopath Dr Leopold Salzer (d. 1907), author of *The Psychic Aspect of Vegetarianism*.[61] Vyvyan quotes Kingsford's 1883 speech welcoming the author and theosophist A.P. Sinnett from India: 'Some of us have dreamed that our English Branch of the Theosophical Society is destined to become the ford across the stream which so long has separated the East from the West, religion from science, heart from mind, and love from learning…'.[62] The same objectives were shared by the Order of the Golden Age and the anti-vivisection movement.

Theosophical, vegetarian and anti-vivisection societies tended, like the OGA, to attract people who had become disenchanted with materialism and scientific 'progress'—it is difficult to imagine them flourishing in pre-industrial Britain—but they were more than just refuges for intellectual refuseniks who yearned for a bucolic utopia that had never existed. They preached a gospel of peace, compassion and spiritual awareness that they hoped would make the new century the beginning of a new age, an age inspired by the Old Testament prophecy of the Messianic Kingdom, and foretold by astrologers as the Age of Aquarius, which was the 'Sign of the Son of Man'.[63] In the years leading up to the Great War, it seemed that the OGA's conciliatory and harmonious ideals might prevail: minor royals, members of the nobility and senior army officers all attended its fundraising concerts, which had a pastoral theme, and enjoyed music and readings extolling the glories of creation, even if they were not sufficiently moved by them to give up meat eating.[64] An unlikely late enthusiast for the Order's objectives of combating 'physical deterioration, disease, and intemperance' was Edward VII, who sent them a message of support as he lay dying in Buckingham Palace.[65]

New Age Politics

Vegetarianism and anti-vivisection did not, of course, fulfil their promise to convert humanity to a more peaceful way of life. The Golden Age never dawned, and the dreams of a peaceable kingdom were shattered by the Great War. Materialism and patriotism became the default positions, and advocates of holistic science, natural living, and international peace were relegated to a marginal counterculture along with dress reformers, naturists, homoeopaths, occultists and sexual liberators, most of whom were linked in the public mind with the politics of Liberalism, Socialism, or even Anarchism, and were thought of at best as eccentrics, and at worst as traitors to their country.[66] According to the *Lancet*, the typical vegetarian was a seditious malcontent who '…cultivates a number of what may be called anti-isms. He is anti-alcoholist, anti-vivisectionist, anti-vaccinationist, anti-capitalist, anti-bellumist, anti-patriotist. He is anti-penalist, and … anti-restraintist, and would abolish all lunatic asylums, rightly from his own point of view, for so he would escape the risk of losing his own liberty'.[67] This was 1916, and a psychiatrist was publicly stating in a leading medical journal that vegetarians and anti-vivisectionists were lunatics who deserved to be locked up.

During the Great War, Oldfield (a pacifist, of course) temporarily abandoned his hospitals to command a casualty clearing station, a service for which he was promoted Lieutenant Colonel and mentioned in dispatches.[68] He was said never to have had a day's illness, but was invalided out of the army in 1918 after being thrown from his horse.[69] He then purchased Margaret Manor near Sittingbourne, which he set up as a fruitarian colony with cottages for adults and communal accommodation for children. According to the advertising, girls were taught fruitarian cookery, and boys, farming, but the children sent there from the slums were left largely unsupervised to roam, and sometimes get lost, in the surrounding countryside.[70] There was little use of medicines: 'epileptics' and 'mental cases' were the commonest types of patients treated there, and those with infectious diseases were banned. Oldfield did, however, take up obstetrics with some success, and acquired an orchard by the characteristically shrewd strategy of inviting all new parents to pay for the planting of a commemorative tree.[71] On Sundays, he would attend divine service in the Manor's private chapel, where Sister Francesca, the mother of his illegitimate daughter, played the organ.[72] In the three years from 1920, Margaret Manor received just £190 in subscriptions, most of which were spent on postage stamps for further charity appeals, but it now had enough long-term residents, who paid up to three guineas a week, to enable Oldfield to close his other establishments and concentrate on running the Manor as a new age retreat.[73]

The OGA attracted fewer members after the First World War, and survived on legacies as ageing spiritualists and animal-loving widows died off—in 1927, for example, Edith Annie Douglas-Hamilton (1871–1927) left £25,000 to the Theosophical Society, £10,000 to anti-vivisection and £5000 to the OGA.[74] By this time, Margaret Manor was far from being the only option for people who wished to pursue a more natural lifestyle. Popular outdoor organizations such as the Scout Movement, the Order of Woodcraft Chivalry, and the Kibbo Kift (archaic Cheshire dialect for 'proof of great strength') all encouraged their members to spend time living in, learning about, and respecting nature. Though this fondness for outdoor pursuits, folkloric traditions and clean living seemed (and sometimes was) the epitome of pastoral innocence, it was not a great step from respecting nature to worshipping it (the rituals of Woodcraft Chivalry and Kibbo Kift influenced those of modern Wicca), or from escaping from capitalist society to rebelling against it.

In 1932, the Kibbo Kift allied itself with the Social Credit movement, a scheme for redistributing wealth, whose founder Major C.H. Douglas (1879–1952) blamed Britain's economic problems on 'international Jewry' and hoped to solve them by paying the British poor for not working. The result was the Green Shirt Movement for Social Credit, an anti-capitalist, anti-government and anti-Semitic group whose aggressive greenness fortunately went no further than minor acts of civil disobedience such as throwing green-painted bricks through government windows.[75] Had things gone their way, they might have started a radical back-to-nature movement in Britain, but no right wing, or green, party ever came close to power. The only European regime officially to endorse natural living, promote spiritual harmony with nature, and ban vivisection, was National Socialist Germany.[76]

For British fascists, hoping to bring in an age of national prosperity by breaking the power of 'international financiers' (for which, read 'Jews'), a 'natural' mode of living was that which corresponded to their own ideology. Even their promise that a natural lifestyle would improve physical and mental health had a dark side, which was that a multitude of problems afflicting the British people, from cancer to criminality, and idiocy to unemployment, could be blamed on malign, and implicitly unnatural, influences such as meat eating, alcohol drinking, and moral and physical degeneracy.[77] In his old age, Oldfield became increasingly concerned that fresh air and cruelty-free living would not be enough to reverse the problem of human degeneration, which could only be confronted by enforcing standards of racial health and purity. In 1944, he wrote in *Healing and the Conquest of Pain,* that '…the crossing of a negro with a white woman is fraught with many curious genetic problems…', and advocated euthanasia for 'idiots'.[78]

While interest in back-to-nature living on the part of British fascists temporarily boosted recruitment, and legacies, to the OGA (one Herbert Jones of Liverpool divided his estate between, among others, the OGA, the RSPCA, the Vegetarian Society, the Malthusian League, and the British Fascisti), it went into terminal decline following the death of Beard in 1938. In Britain on the eve of war, the Order's fascist links were a humiliating liability, and it decamped to South Africa, where it survived until 1959.[79] The continuation of the humane movement in the post-war period will be the subject of chapter seven, but we will first consider the successor to Oldfield's anti-vivisection hospital, and the medical profession's response to it.

Notes

1. J. Howard Moore, 'Universal kinship', *Herald of the Golden Age*, 11 (1906), 38–42 (Moore 1906).
2. *Vegetarian Advocate*, March 1850, cited in Kathryn Gleadle, '"The age of physiological reformers:" rethinking gender and domesticity in the

age of reform', in Arthur Burns and Joanna Innes (eds) *Rethinking the Age of Reform: Britain 1780–1850* (Cambridge: Cambridge University Press, 2003), 213. Vegetarianism based on Hinduism had been promoted in print as long ago as 1791, in John Oswald's *The Cry of Nature*. Some early vegetarians eschewed all animal products and others, 'fruitarians', avoided killing living things (Gleadle 2003).

3. Henry Stephens Salt, *Animals' Rights Considered in Relation to Social Progress* (New York, Macmillan, 1894), 8 (Salt 1894).
4. Stuart, *Bloodless Revolution*, 422–423. In 1850, 16 of 478 Vegetarian Society members were medics: Julia Twigg, *The Vegetarian Movement in England: 1847–1981: A Study in the Structure of its Ideology*. PhD, London School of Economics (1981), 87 (Twigg 1981).
5. Dellar, *Josiah Oldfield*, 19.
6. Josiah Oldfield. *A Groaning Creation* (London: Ideal Publishing Union, n.d.); *A Tale of Shame and Cruelty* (Paignton: Order of the Golden Age, n.d.) (Oldfield n.d.; A Tale of Shame and Cruelty n.d.).
7. 'A new guild', *Dundee Courier*, 27 July 1882, 3 (A New Guild 1882).
8. 'Vegetarian banquet', *Bristol Mercury*, 30 March 1883, 6; editorial, *Herald of the Golden Age*, *12* (1908), 12 (Vegetarian Banquet 1883).
9. *Herald of the Golden Age*, *10* (1903), inside front cover.
10. 'General News', *Derby Daily Telegraph*, 27 January 1883, 4 (General News 1883).
11. Editorial, *Aberdeen Evening Express*, 17 November 1883, 2 (Editorial 1883).
12. Josef Francis Charles Craven, *Redskins in Epping Forest: John Hargrave, the Kibbo Kift and the woodcraft experience*. PhD, University College London (1998), 13; *The Order of the Golden Age* [flyer], n.p., [1904], 2, http://www.ordergoldenage.co.uk/page34.html, viewed 20 June 2016 (Craven 1998).
13. Josiah Oldfield, *The Claims of Common Life, or, The Scientific Relations of Humans and non-Humans* (London: Ideal Publishing Union, 1898), 17 (Oldfield 1898).
14. 'A vegetarian hospital', *Herald of the Golden Age*, *1* (1896), 150 (A Vegetarian Hospital 1896).
15. The Order of the Golden Age taught that cancer was curable by a vegan diet: 'A dietetic cure of cancer' (editorial), *Herald of the Golden Age*, *8* (1903), 114 (A Dietetic Cure of Cancer 1903).
16. Oriolet Hospital, *Annual Report* (1896), 9–11.
17. 'A medical sack race', *BMJ*, *2* (1897), 1115–1116.
18. 'Ireland', *Times*, 29 January 1898, 12 (Ireland 1898).

19. London Metropolitan Archives (hereinafter LMA) A/FWA/C/D330/1; *Morning Star*, 23 March 1897.

20. Flyer appealing for 'Hospital of St. Francis', April 1897, LMA A/FWA/C/D330/1.

21. *A Tale of Shame*, back cover.

22. 'Hospital of St. Francis', LMA A/FWA/C/D330/1. Anti-vivisectionists made much of the supposed link between vivisection and experiments on the poor: see, for example, Mark Thornhill, *Experiments on Hospital Patients* (London: Hatchards, 1889) (Thornhill 1889).

23. Statement by Charles Carthew, 21 February 1898, LMA A/FWA/C/D330/1.

24. Oldfield was born at Ryton, Shropshire, on 28 February 1863 and graduated BA in theology in 1885. He then trained as a barrister at Lincoln's Inn and practised on the Oxford circuit. For a fascinating history of his life, see Rosemary Dellar, *Josiah Oldfield: Eminent Fruitarian* (Raynham: Rainmore Books, 2008) (Dellar 2008).

25. Letter, COS to Rev'd A.G. Deedes, 20 May 1898, LMA A/FWA/C/D330/1.

26. Letter, W. Vaughan Morgan to COS, 8 December 1898, LMA A/FWA/C/D330/1.

27. COS report, 25 April 1898, LMA A/FWA/C/D330/1.

28. *Medical Directory* (1900), 459.

29. Dellar, *Josiah Oldfield*, 97 (2008) (Dellar 2008).

30. 'The institutional workshop: the Hospital of St Francis', *The Hospital*, *32* (1902), 296–7.

31. 'Hospital Sunday Fund', *Times*, 17 June 1902, 8 (Hospital Sunday Fund 1902).

32. Dellar, *Josiah Oldfield*, 96.

33. COS memorandum, 7 April 1904, LMA A/FWA/C/D330/1.

34. Flyer for Lady Margaret Hospital, reprinted in *South London Press*, 29 August 1904, LMA A/FWA/C/D330/1 (Flyer for Lady Margaret Hospital 1904).

35. 'Lady Margaret Hospital', *Herald of the Golden Age*, *12* (1908), 14; 'The Fruitarian Hospital', *Herald of the Golden Age*, *11* (1906), 35 (Lady Margaret Hospital 1908; The Fruitarian Hospital 1906).

36. 'Brother pain and his crown' (editorial), *Herald of the Golden Age*, *11* (1906), 14; Anon., letter to COS, n.d., LMA A/FWA/C/D330/1 (Brother Pain and His Crown 1906).

37. Note by Hugh Fickling, 25 April 1910, LMA A/FWA/C/D330/1.
38. Anon., 'The Lady Margaret Hospital, Bromley', *Lancet*, *1* (1906), 1088. He was equally controversial in promoting his legal practice: Dellar, *Josiah Oldfield*, 55–56 (Anon 1906).
39. Letter, M. Michael to E.S. Kemp, 13 August 1913; COS notes on Margaret Manor, LMA A/FWA/C/D330/1.
40. COS notes, 22 April 1908, LMA A/FWA/C/D330/1.
41. Vyvyan, *In Pity*, 133–4 (Vyvyan 1969).
42. Killick Millard, 'The rôle of the "anti": an apology and an appeal', *Public Health*, *15* (1902), 212–222 (Millard 1902).
43. Anon., *The Order of the Golden Age: Its Aims, Its Objects and Its Rules* (n.p., 1904), 1 (Anon 1904).
44. Sidney H. Beard, 'The festival of the Christians', *Herald of the Golden Age*, *8* (1903), 133–134 (Beard 1903).
45. Josiah Oldfield, 'The Christmas feast: an indictment', *Evening News*, 11 December 1912 (Oldfield 1912).
46. Beard, 'The festival of the Christians'. The moral opposition of flesh and spirit was influenced by Pauline theology, e.g., Romans 8.
47. Robert H. Perks, *Why I Condemn Vivisection* (Paignton: OGA, 1992) (Perks 1992).
48. 'The vivisection libel suit', *Herald of the Golden Age*, *8* (1903), 190: 'funds' refers to state-controled charities such as the King's Fund (The Vivisection Libel Suit 1903).
49. *Herald of the Golden Age*, *passim*, inside back cover.
50. S.L. Gill, 'Case of an opium-eater and vegetarian becoming bedridden: recovery on taking animal food', *Lancet*, *2* (1853), 95 (Gill 1853).
51. 'Vegetarianism and physique' (editorial), *Lancet*, *2* (1908), 1537.
52. Craven, *Redskins in Epping Forest*, 113 (1998) (Craven 1998).
53. Isaiah 11: 6–9.
54. Flyer for Hospital of St Francis, 1897, LMA A/FWA/C/D330/1.
55. Josiah Oldfield, *Claims of Common Life*, 7–13.
56. 'From "Realization"', *Herald of the Golden Age*, *8* (1903), 119; Moore, 'Universal kinship'.
57. Oldfield, *Claims of Common Life*, 49, 51, 54, 70–72.
58. Edward E. Lond, 'A world religion', *Herald of the Golden Age*, *11* (1906), 20–221. Ecumenism had received a boost from the 1893 *World's Parliament of Religions* in Chicago (Lond 1906).

59. Kate Cording, 'A talk with the children', *Herald of the Golden Age*, *8* (1903), 137; *The Order of the Golden Age*, 3; Sidney H. Beard, 'Our national peril', *Herald of the Golden Age*, *10* (1903), 108–111 (Cording 1903; Beard 1903).

60. She wrote, 'I have killed Paul Bert, as I killed Claude Bernard; as I will kill Louis Pasteur, and after him the whole tribe of vivisectors … it is a magnificent power to have, and one that transcends all vulgar methods of dealing out justice to tyrants': Alan Pert, *Red Cactus: the Life of Anna Kingsford* (Watsons Bay New South Wales: Books and Writers, 2007), 200 (Pert 2007).

61. James R.T.E. Gregory, '"A Lutheranism of the table": religion and the Victorian vegetarians', in Rachel Muers and David Grumett (eds), *Eating and Believing: Interdisciplinary Perspectives on Vegetarianism and Theology* (London: T. and T. Clark, 2008), 135–151 (Gregory 2008).

62. Vyvyan, *In Pity*, 140, 145.

63. Nina A. Hutteman Hume, 'The higher aspects of the simple life', *Herald of the Golden Age*, *12* (1908), 6–9; Sidney H. Beard, 'The prevention of pain', *Herald of the Golden Age*, *12* (1908), 1–3 (Hume 1908; Beard 1908).

64. 'Order of the Golden Age', *Times*, 1 November 1910, 16 (Order of the Golden Age 1910).

65. 'Court Circular', *Times*, 5 May 1910, 13. A formal expression of support from the sovereign did not necessarily indicate personal interest, but it is significant that the OGA was deemed worthy of royal endorsement (Court Circular 1910).

66. Craven, *Redskins in Epping Forest*, 43–44 (1998) (Craven 1998).

67. Chas. Mercier, 'Diet as a factor in the causation of mental disease', *Lancet*, *1* (1916), 565 (Mercier 1916).

68. Virginia Smith, 'Oldfield, Josiah (1863–1953)', *Oxford Dictionary of National Biography* (Oxford: Oxford University Press, 2004), http://www.oxforddnb.com/view/article/40999, viewed 20 June 2016 (Smith 2004).

69. 'Ashford', *Whitstable Times and Herne Bay Herald*, 25 May 1918, 6 (Ashford 1918).

70. M.J. Ferrers to the COS, 19 November 1919, LMA FWA/C/D330/1.

71. *Truth*, 26 May 1926, cutting in LMA FWA/C/D330/1.

72. COS report, September 1913, LMA A/FWA/C/D330/1.

73. 'An obnoxious nursing home' (editorial), *Truth*, 30 December 1925, 1221–1222. One resident was the editor of the Journal of the League for the Abolition of Cruel Sports, Henry Brown Amos (1869–1946) (An Obnoxious Nursing Home 1925).

74. 'Woman millionaire's bequests', *Fife Free Press*, 15 October 1927, 9 (Woman Millionaire's Bequests 1927).

75. According to the Greenshirt newsletter, *Attack!*, published in the early 1930s, they aimed to break the power of 'a handful of power-mad money lenders': 'We attack the bankers!', *Attack!, 9* (1933), 1.

76. On the links between environmentalism and Nazism, see Anna Bamwell, *Blood and Soil: Richard Walter Darre and Hitler's 'Green Party'* (Buckinghamshire: Kensall Press, 1985); *Ecology in the 20th Century: A History* (New Haven: Yale University Press, 1989) and *The Fading of the Greens* (New Haven: Yale University Press, 1994) (Bamwell 1985, 1989, 1994).

77. Alex Haag, 'The true cause of physical degeneracy', *Herald of the Golden Age, 11* (1906), 73 (Haag 1906).

78. Dellar, *Josiah Oldfield*, 299–300. Oldfield continued to be active until well over the age of eighty. Though he unfortunately fell out of a rocking chair and broke his leg during a trip to the West Indies, he did not lose a chance to promote vegetarianism, telling the press that 'Since all of nature's common diseases to kill people off are powerless against fruitarianists, she takes a mean advantage of fruitarian veterans by increasing the fragility of old people's bones, so that even a fruitarian runs a danger of a broken bone if he frolics on polished staircases or marble floors or goes mountain climbing', and giving his age as 97: 'Nature's "mean trick" on oldest doctor', *Hull Daily Mail*, 11 March 1950, 5.

79. 'Deaths', *Times*, 21 October 1938, 1; Wills and bequests', *Times*, 28 January 1937, 10 (Deaths 1938; Wills and bequests 1937).

References

A Dietetic Cure of Cancer (Editorial). (1903). *Herald of the Golden Age, 8*, 114.

A New Guild. (1882, July 27). *Dundee Courier, 3*.

A Vegetarian Hospital. (1896). *Herald of the Golden Age, 1*, 150.

An Obnoxious Nursing Home (Editorial). (1925, December 30). *Truth*, 1221–1222.

Anon. (1904). *The Order of the Golden Age: Its Aims, Its Objects and Its Rules* (n. p.), 1.

Anon. (1906). The lady Margaret hospital Bromley. *Lancet, 1*, 1088.

Ashford. (1918, May). *Whitstable Times and Herne Bay Herald, 25*, 6.

Bamwell, A. (1985). *Blood and soil: Richard walter darre and Hitler's 'green party'*. Buckinghamshire: Kensall Press.

Bamwell, A. (1989). *Ecology in the twentieth century: A history*. New Haven: Yale University Press.

Bamwell, A. (1994). *The fading of the greens*. New Haven: Yale University Press.

Beard, S. H. (1903a). The festival of the Christians. *Herald of the Golden Age, 8*, 133–134.

Beard, S. H. (1903b). Our national peril. *Herald of the Golden Age, 10*, 108–111.

Beard, S. H. (1908). The prevention of pain. *Herald of the Golden Age, 12*, 1–3.

Brother Pain and His Crown. (Editorial). (1906). *Herald of the Golden Age, 11*, 14.

Cording, K. (1903). A talk with the children. *Herald of the Golden Age, 8*, 137.

Court Circular. (1910, May 5). *Times, 5*, 13.

Craven, J. F. C. (1998). *Redskins in Epping Forest: John Hargrave, the Kibbo Kift and the woodcraft experience*. Ph.D, University of London.

Deaths. (1938, October 21). Times, 1.

Dellar, R. (2008). *Josiah Oldfield: Eminent fruitarian*. Raynham: Rainmore Books.

Editorial. (1883, November 17). *Aberdeen Evening Express*, 2.

Flyer for Lady Margaret Hospital. (1904, August). Reprinted in *South London Press*, 29.

General News. (1883, January 27). *Derby Daily Telegraph*, 4.

Gill, S. L. (1853). Case of an Opium-Eater and vegetarian Beoming Bedridden: Recovery upon Taking Animal Food. *The Lancet, 62*(1561), 95.

Gleadle, K. (2003). "The age of physiological reformers:" Rethinking gender and domesticity in the age of reform. In A. Burns & J. Innes (Eds.), *Rethinking the age of reform: Britain 1780–1850* (p. 213). Cambridge: Cambridge University Press.

Gregory, J. R. (2008). "A Lutheranism of the table": religion and the Victorian vegetarians. (2008). In R. Muers, & D. Grumett (Eds.), *Eating and believing: Interdisciplinary perspectives on vegetarianism and theology*. London: T & T Clark.

Haag. A. (1906). The true cause of physical degeneracy. *Herald of the Golden Age, 11*.

Hospital Sunday Fund. (1902, June 17). *Times,* 8.

Hume, N. A. H. (1908). The higher aspects of the simple life. *Herald of the Golden Age, 12,* 6–9.

Ireland. (1898, January 29). *Times,* 12.

Lady Margaret Hospital. (1908). *Herald of the Golden Age, 12,* 14.

Lond, E. E. (1906). A world religion. *Herald of the Golden Age, 11,* 20–22.

Mercier, C. (1916). Diet as a factor in the causation of mental disease. *The Lancet, 187*(4827), 510–513.

Millard, C. K. (1902). The rôle of the "anti": An apology and an appeal. *Public Health, 15,* 212–222.

Moore, J. H. (1906). Universal kinship. *Herald of the Golden Age, 11,* 38–42.

Oldfield, J. (1898). *The claims of common life, or, the scientific relations of humans and non-humans.* London: Ideal Publishing Union.

Oldfield, J. (1912, December 11). The Christmas feast: An indictment. *Evening News,* 11.

Oldfield, J. (n.d.). *A groaning creation.* London: Ideal Publishing Union.

Oldfield, J. (n.d.). *A tale of shame and cruelty.* Paignton: Order of the Golden Age.

Order of the Golden Age. (1910, November 1). *Times,* 16.

Perks, R. H. (1992). *Why i condemn vivisection.* Order of the Cross. Paignton.

Pert, A. (2007). *Red cactus: The life of Anna Kingsford.* Watsons Bay: Books and Writers.

Salt, H. S. (1894). *Animals' rights considered in relation to social progress.* New York: Macmillan.

Smith, V. (2004). Oldfield, Josiah (1863–1953). *Oxford Dictionary of National Biography.* Oxford: Oxford University Press. Retrieved June 20, 2016, from http://www.oxforddnb.com/view/article/40999.

The Fruitarian Hospital. (1906). *Herald of the Golden Age, 11,* 35.

The Vivisection Libel Suit. (1903). *Herald of the Golden Age, 8,* 190.

Thornhill, M. (1889). *Experiments on hospital patients.* London: Hatchards.

Twigg, J. (1981). *The vegetarian movement in England: 1847–1981: A study in the structure of its ideology.* PhD, London School of Economics.

Vegetarian Banquet. (1883, March 30). *Bristol Mercury,* 6.

Vyvyan, J. (1969). *In Pity and in Anger.* London: Michael Joseph.

Wills and bequests. (1937, January 28). *Times,* 10.

Woman Millionaire's Bequests. (1927, October 15). *Fife Free Press,* 9.

5

The National Anti-Vivisection Hospital, 1902–1935

What exactly was an anti-vivisection hospital? The question arose at an inquest into the death of a child, Mabel Florence Jones, in 1908. She had been treated at the National Anti-Vivisection Hospital in Battersea, South London, for 'a clean-cut wound in her head'; a few weeks later, she suddenly became ill and was taken to St Bartholomew's Hospital, where it was discovered that her skull was fractured. According to the house surgeon at St Bart's, the fracture would have been discovered sooner if the child had been 'properly examined' at Battersea, but Dr Ronald da Costa, the Anti-Vivisection Hospital's resident medical officer, told the court he had examined the child thoroughly and that the fracture must have been the result of 'a subsequent accident'. The background to the case suggests parental neglect: the child had been dirty when taken to hospital, and at Bart's the surgeon had found it 'extremely difficult to get a history' from the mother, who first said that Mabel had been hit by a pickaxe carelessly thrown over a wall, and then that she had 'knocked her head against the table'. Unfortunately, poor record-keeping left the Anti-Vivisection Hospital struggling to defend itself. Dr da Costa, a former Indian Army surgeon with 30 years' experience, admitted his hands had been 'pretty full', as in addition to the hospital's in-patients, he attended out-patients from six-thirty until

© The Author(s) 2017
A.W.H. Bates, *Anti-Vivisection and the Profession of Medicine in Britain*,
The Palgrave Macmillan Animal Ethics Series, DOI 10.1057/978-1-137-55697-4_5

eleven in the evening, seeing scores of patients who paid four pence each. Battersea, the court heard, was one of the poorest suburbs of London, and in Henley Street, where the Joneses lived, slum-dwellers were 'huddled together like sheep'.[1]

The Coroner tried to clarify the objects of the hospital that was treating so many poor South Londoners: 'I see in your hospital's annual report the words, "No experiments on patients". What do they mean?' he asked da Costa, who replied: 'I don't know. I don't suppose any hospital has experiments on patients. We perform every kind of operation'. The hospital's secretary (chief administrator), G.W.F. Robbins, explained that it meant no interventions for the sake of knowledge rather than the benefit of patients. To the Coroner's suggestion that this implied other hospitals *did* experiment, Robbins answered 'Not in the least'.[2] It seemed, however, that local people thought otherwise: one of the medical men present had heard that some people were afraid to go into hospital in case they were experimented on. The accounts of 'human vivisection' that sometimes appeared in British newspapers referred to therapeutic experiments in European hospitals rather than patients being dissected alive in Britain, but they probably stoked fears that London's medical men did not willingly confine their researches to animals.[3] The Anti-Vivisection Hospital, by requiring its staff to sign a personal pledge not to experiment on animals or humans, was deliberately presenting itself as a safer alternative to Bart's and other London teaching hospitals where vivisectionists worked.

Thanks to the indefatigable Stephen Coleridge, everybody thought they knew where London's vivisectionists were. In 1901, he had published a list, *The Metropolitan Hospitals and Vivisection*, with a 'blood-red band' of 'fearsome appearance' against the names of the offending hospitals and schools. This caused serious concern to London's teaching hospitals, which feared that the 'stream of charity' might be diverted away from them if donors thought they were paying for vivisection, and there ensued a convoluted public argument over how much, if any, charitable giving actually found its way into the laboratories of vivisectionists. To add insult to financial injury, Coleridge used the familiar anti-vivisection argument from the nineteenth century—that vivisectionists were, or became, callous and were thus unfit to attend patients—to imply that teaching hospital staff were blasé about using their patients for research: the London School of Tropical Medicine's boast of 'an adequate supply of clinical material for purposes of instruction' was a circumlocution, he told his readers, for 'the prostrate bodies of the sick'.[4] He knew that money was being raised for a new hospital that would promise freedom from experimentation, and to which donors would be able to give with the assurance that they were not supporting animal research.

Josiah Oldfield's tiny, shabby, hastily conceived Hospital of St Francis (discussed in the previous chapter) was, though ground-breaking, scarcely a flagship for the cause, and Coleridge, its chairman, had plans to open a larger hospital once the money had been raised to buy suitable premises. St Francis's had at least shown that it was possible for an anti-vivisection hospital to survive without the support of the medical establishment. Furthermore, it seemed to have caused them some anxiety: the great Sir Henry Burdett had taken the trouble to attack Oldfield's 'little hospital in the New Kent Road run on the cheap and simple plan of giving its patients no meat', professing to find it 'somewhat amusing … that Mr Coleridge, who poses as an opponent of experiments on animals, should be the chairman of a vegetarian hospital which exists for the very purpose of trying a very doubtful experiment on human beings'. Burdett's financial acumen and blustering philanthropy were invaluable to the voluntary hospital system, and he was chary of anyone or anything that threatened its funding, particularly '[t]hose peculiar people who have chosen to associate themselves with the National Anti-Vivisection Society [of which Coleridge was secretary] and to set themselves in opposition to such great charitable movements as the [Metropolitan] Hospital Sunday Fund and King Edward's Hospital Fund'.[5]

The NAVS had been raising money for a hospital since the 1890s, when it set up a charitable trust with Lords Coleridge (1820–1894, Stephen's father) and Hatherton (1842–1930), Dr Abiathar Wall (treasurer of the London Anti-Vivisection Society), Ernest Bell (1851–1933, later president of the Vegetarian Society and chairman of the NAVS) and the Rev'd Augustus Jackson as trustees. Helped by rich anti-vivisectionists (particularly the Dowager Countess of Portsmouth, who replaced Coleridge as a trustee), who donated as an alternative to supporting the 'vivisecting' teaching hospitals such as Bart's, Guy's, the London, Mary's and St Thomas's, the fund grew rapidly.[6] As usual among anti-vivisectionists, there were personality clashes—according to the hospital's first medical director, Dr Alexander Bowie, 'there were two sections who differed about it for a time and the money went into Chancery'—but in 1900 they agreed to buy, for seven thousand pounds, a large private house, known as Lock's Folly, in Battersea, where, two years later, an out-patient department was opened in the former stables.[7]

In-patients followed in 1903: there were 'eight beds, three cots, and four medical officers, one of whom is also chairman of the hospital'. Its prospectus promised 'No Vivisection in its Schools. No Vivisectors on its Staff. No Experiments on Patients', and listed eleven 'honorary' medical staff who, as in all voluntary hospitals, were consultants who did not receive salaries but had the kudos of being associated with a London hospital, albeit in this case an obscure one, which supposedly helped their private practice.[8] In common with most voluntary hospitals, the Anti-Vivisection charged patients a small fee—which might have been paid by one of the hospital's subscribers, an employer, or a provident association—but the bulk of the running costs were met by donations. South London was notoriously short of hospital beds and the local poor, who made up ninety per cent of the new hospital's patients, had few alternatives; nevertheless, Robbins boasted that '[t]he special effect of our principles upon the sick poor is to attract them to our Hospital in which they seem to have complete confidence, as is proved by the rapid increase in our work'.[9]

In an effort to entice patients, and donors, away from other institutions, the hospital spent some three hundred pounds a year advertising in anti-vivisection periodicals.[10] Its annual report described it as 'a standing protest against cruel experiments on animals, and a concrete demonstration that these are not necessary for the succour of the maimed or the healing of the sick'. Its patients were reassured that 'the whole of the medical surgical and administrative staff are pledged against vivisection', that treatments such as Pasteur's vaccines that were prepared using live animals were 'absolutely shut out', and that, except in an emergency, no operation would be performed without the written consent of the patient, a novel undertaking for the time, though no records survive to show whether or not it was adhered to.[11]

The Anti-Vivisection Hospital and the King's Fund

In 1906, Lord Lichfield (1856–1918) and the surgeon Sir William H. Bennett (1852–1931) visited the Hospital on behalf of King Edward's Hospital Fund (the King's Fund). They noted the 'steadily increasing amount of work done', which showed that 'this Institution supplies a want', and found it tolerably well equipped, with modern electrical fittings, though they thought the medical staff were 'gentlemen who are hardly of the status of those who usually occupy positions on the staffs of Hospitals of repute in London'. They submitted a detailed report to the Fund, but the outcome was apparently a foregone conclusion, and the Fund ruled that further visits would be 'a thankless task, which can be productive of no good', since, 'the anti-vivisection basis upon which the Institution is founded is in itself considered a sufficient reason for withholding help'.[12]

The King's Fund was one of three funds that distributed public donations to London's voluntary hospitals, the others being the Sunday and Saturday Funds, which administered church and workplace collections, respectively.[13] The King's Fund was the largest of the three, controlling an £140,000 annual distribution, which gave it great influence over practices and standards. It refused to tell the Anti-Vivisection Hospital why it had refused it a grant, but Sydney Holland (1855–1931), an outstanding fundraiser, founder of the Research Defence Society (see Chap. 6), and chairman of the London Hospital (one of the King's Fund's biggest beneficiaries), was more forthcoming. At a charity dinner in 1908, he called the Anti-Vivisection 'a miserable hospital, miserably built and miserably equipped' (perhaps not the soundest of reasons for *not* giving it money), and he later told Robbins that both the King's Fund and the Sunday Fund were refusing to help because it was unfit, as a building, to be a hospital.[14] This was wrong both in fact—the hospital continued in the same premises until 1972—and as an explanation for the King's Fund's decision, and it appears that Holland, whose hospital had been among those accused of using charitable donations to pay for experiments on animals, was attempting to discourage him from seeking funds in future. The Anti-Vivisection Hospital's continued existence was, as it was intended to be, an irritation to the funds and the hospitals they supported: at the Sunday Fund's annual meeting in the Mansion House (the official residence of the Lord Mayor of London) the self-congratulatory speeches were interrupted by the Rev'd Lionel Lewis (1867–1953), Vicar of St Mark's Whitechapel, complaining, not for the first time, that the Anti-Vivisection had not received a grant. When the former Lord Mayor Sir John Bell (1843–1924), a wealthy brewer, responded that he would not bandy words with someone whose church had only contributed 3s 6d the Fund, the vicar replied, 'Yes, my parish is a poor one'.[15] This vignette shows something of the patrician distain for anti-vivisectionists among the philanthropists of the City of London, whose liverymen controlled much of the capital's charity.

The conservative *Medical Times* defended the Anti-Vivisection, which it described as 'a small hospital for the treatment of the indigent sick', against Holland's criticisms, on the grounds that 'Patients are not bound to go to the hospital...', and noted that '[t]here are lots of hospitals in London doing excellent work which are 'miserably equipped', adding, 'we are unable to understand why such a dead set has been made against the hospital. Enormous influence is being brought to bear to crush this modest attempt ... with such success that the Distribution Committees of the great Hospital Funds in London have been induced to refuse ... funds'. The *Medical Times* proposed that, as with the Homoeopathic and Temperance hospitals, the experiment should be allowed to proceed, the result being determined by whether the hospital proved viable or not.[16]

The management of the Anti-Vivisection Hospital seemed genuinely unable to understand why the funds had refused them support, and continued to solicit inspections in the hope of getting the decision reversed. In the 1890s, when the hospital was being planned, it would have been reasonable to have anticipated that the Prince of Wales's Fund (as the King's Fund then was), instituted to commemorate the jubilee of Queen Victoria, would be sympathetic towards anti-vivisection; indeed, had the hospital opened during Victoria's lifetime, she may well have become its patron. Unfortunately, neither the future Edward VII nor the future George V, who as Princes of Wales chaired the King's fund, shared Victoria's anti-vivisection principles, and the medical men who made up the Fund's committee all sympathised with vivisection even if they did not, like its chairman Lord Lister (1827–1927), practise it themselves.[17]

At a public meeting to discuss the Hospital's funding, Stephen Coleridge, an agreeable man in private life, was characteristically forthright: 'these great funds had got into the hands of persons who deliberately disposed of them to forward the views they personally held'.[18] However, since the Anti-Vivisection Hospital had initially been put up as an alternative to supporting the funds, to allow anti-vivisection donors who wanted to give to a hospital to do so directly without 'sacrificing their consciences', its trying to claim from the funds as well looked like an attempt to have one's cake and eat it, and it is not surprising that the funds were indignant.[19]

The funds were, however, supposed to be impartial, and could not risk openly stating they were on the side of vivisection, so the Sunday Fund disingenuously said they had refused the Anti-Vivisection a grant because 'the best treatment known' (presumably antisera) was not given and the 'Governing Body dictates the forms of treatment to be used by its medical staff'.[20] However, the latter was true of most of the voluntary hospitals supported by the fund, including the Homoeopathic and Temperance, and the National Anti-Vivisection actually had more medical board members than most, three doctors and a dentist in 1910, though these were all committed anti-vivisectionists like Lizzy Lind af Hageby (1878–1963), founder of the Animal Defence and Anti-Vivisection Society (ADAVS), whose infiltration of the laboratories at University College London in 1903 had paved the way for the notorious Brown Dog affair. The prominence of women in anti-vivisection circles meant that they too were well represented (five out of eighteen) on the hospital's board.[21]

The Hospital informed the Sunday Fund that its rules banning vivisection and treatments derived from it were no different 'to those which in a Homoeopathic Hospital prevent the appointment of Allopaths, and in a Temperance Hospital the appointment of alcohol drinkers, and which prohibit any alcoholic treatment'.[22] The Sunday Fund, however, said the difference was that patients understood what a homoeopathic hospital was, but not an anti-vivisection one,[23] to which the Hospital's committee replied: 'the poorer class in Battersea, who appreciate our Hospital … well understand its title. Possibly the statue of the dog erected there conduces to this knowledge'.[24]

The reference was to an episode that the people of Battersea would have found hard to ignore: the Brown Dog affair (1907–1909), an anti-vivisection *cause célèbre* that prompted rioting on the streets as medical students attempted to demolish Battersea Council's provocative—the British Medical Association thought libellous—bronze statue of a terrier allegedly 'done to death' in the laboratories of University College. Battersea's menfolk came out in force to defend the dog and 'guard' the hospital; one perhaps rather fanciful account of the riots has them forcibly preventing an injured medical student—whose objection to the hospital had conveniently evaporated when he needed its help—from passing through its doors.[25] These 'town against gown' fights were clearly about more than just vivisection: they channelled the 'pent-up hatred felt by certain classes towards medical science and medical men', who, 'brutalised by vivisection', would not hesitate to experiment on the poor. The students' reaction to hearing that their efforts to learn their trade were brutalising them was, with presumably unintended irony, to start a riot.[26] Whatever the rights and wrongs of the disturbances, the Sunday Fund's claim that potential patients did not know about anti-vivisection was incredible; Battersea people would have seen and heard the riots, they were widely reported in London's papers, and the dog's statue close by the hospital served as a constant reminder.

Sentiment and Science

Some commentators have seen the Brown Dog affair as damaging to anti-vivisection because it overtly politicised it. By defending the dog, Battersea's poor were fighting for more than their right not to be experimented on without their consent; they were using the fate of the dog as a symbol for the exploitation of working class men and women in ways that had nothing to do with vivisection.[27] Medical students were a particular target of the workers' animosity because they represented the sort of unsympathetic middle class parvenus who, unlike the aristocracy, could not be relied upon to treat those who came under their authority with any decency. The students' behaviour, which was tolerated, or even encouraged, by their professors, only reinforced these concerns: rowdyism in defence of science suggested that doctors who insisted upon experimenting on animals were indeed uncaring. One anti-vivisectionist who was present at a meeting disrupted by students shouting and breaking up the furniture wrote to the King's Fund: 'the disgraceful scene at Caxton Hall made by the medical students fills one with the utmost contempt and loathing for these cowardly "butchers" … if these creatures are our coming medical men, then I say—God help their unfortunate patients!'[28] 'Butchers' was a common slur against unpopular medics, since butchers were proverbial for their lack of compassion (and thus reputedly banned from serving on juries), and to compare medics to them, as Lind af Hageby did in *The Shambles of Science*, was to deny medicine its claim to moral and ethical superiority.

The riots show that the medical exponents of vivisection were indifferent to public expectations that they ought to approach their work with compassion, and failed to take seriously concerns that there was a slippery slope from vivisection to experiments on patients. In their crudest form, these concerns were, indeed, no more than fantasies fed by popular fiction, in which mad surgeons with 'elastic' consciences, their appetites for cruelty 'whetted' by experimenting on animals, easily overcame their 'natural prejudice against inflicting suffering' and took to vivisecting humans.[29] However, not only impressionable folk who took such tales literally (they are reminiscent of the mid-nineteenth century tales of paupers dissected and made into 'anatomy pie') would have felt safer in the Anti-Vivisection Hospital, but also those who thought that medics prepared to sign an anti-cruelty pledge would probably treat their patients more sympathetically than most.

To be anti-vivisection was to reject the cold, rational, science on which the teaching hospitals based their reputation for excellence, and be guided by emotion: Lind af Hageby claimed she converted to the cause after the look of suffering on the face of a laboratory dog went, as she put it, 'straight to my heart'. Medical men (and they were always men) complained about womanish emotionality (the American medical profession accused anti-vivisectionists of being literally mad with a mental condition called 'zoophil-psychosis'[30]) but potential patients seem to have agreed with the National Anti-Vivisection Society that for a doctor to be called sentimental was an accolade.[31]

To sign the pledge made, after all, no practical difference to animals—only a tiny minority of medical practitioners ever conducted laboratory experiments, and no busy district hospital had time for them—its purpose was to affirm the commitment of the Hospital's staff to compassionate medicine, and it would have reassured not just patients with strong views on animal experimentation, but also those concerned that charitable hospitals were no longer the benevolent institutions they had once been, but testing-grounds, run for the convenience of an increasingly powerful and self-interested medical profession.[32]

Even though it helped attract patients, for a doctor to sign the anti-vivisection pledge was professionally hazardous: an act of public dissent from the dogmas of the metropolitan medical élite that marked one out as a medical protestant. It was rumoured that 'no one with pronounced anti-vivisectional principles would be elected to the Medical Staff of the larger London Hospitals', and '[t]he retention of his appointment by anyone opposing vivisectional teaching would be difficult, if not impossible and his promotion unlikely'.[33] In view of this, it seems that some, perhaps most, medics who had qualms about vivisection did not voice them. As Walter Hadwen (1854–1932), a Gloucester GP who took over from Cobbe as president of the British Union for the Abolition of Vivisection (BUAV), told an audience of anti-vivisectionists in 1907, anyone prepared to made a stand 'had to learn what it meant to be heterodox',[34] and not all had the courage of their convictions: one anti-vivisection doctor described how 'A young Doctor told a lady that he hated Vivisection, but did not dare express it, or he would have been hooted out of the Profession'.[35]

The British Medical Association, the doctors' trade union, treated anti-vivisectionists as enemies of the profession and smugly observed each year in its journal that the Anti-Vivisection Hospital had once more received nothing from the funds. As the size of each hospital's grant was roughly proportionate to the number of patients seen and the cost of any new buildings, it served as an indicator of a hospital's achievements, so the persistent withholding of an award might embarrass a hospital even if it did not bother it financially. In 1908, the Anti-Vivisection received a one-off payment of £440 from the Sunday Fund, given, according to Burdett, in the hope that 'the [hospital] authorities would mend their ways, purge their methods, and, in fact, fall into line…' It also gave him another opportunity to state in his journal that it was 'pretentious humbug' to employ modern methods that had been developed from experiments on animals and then boast of an anti-vivisection commitment, in response to which, Robbins challenged him to name a treatment whose development *had* depended on vivisection, an invitation to an argument that, though popular over the years with both pro- and anti-vivisectionists, was always unproductive, since even if it were agreed that a particular medical advance had resulted from vivisection (which it never was), it was practically impossible to prove that it could *only* have been made in this way and not by some other means.[36] Burdett also repeated the accusation he had previously made about the Hospital of St Francis, that 'No experiments on hospital patients' was nonsense since 'the whole hospital is an experiment', though he seemed disinclined to wait, as the *Medical Times* had suggested, for its outcome.

The King's Fund's concern that 'anti-vivisection propaganda' would lead to a drop in their income appears to have been justified: one irate correspondent cited 'the devilish abomination called vivisection' as the reason that '[m]y sister and I, together with several of our friends, therefore, do not intend to give one farthing to any other hospital; all the money we can spare we will send to the Battersea Hospital', adding, [w]e also believe that where hospitals practise vivisection upon helpless animals they, naturally, will not stop there, it is not reasonable to expect them to do so'.[37] In 1907, the Fund's committee, with the Prince of Wales in the chair, had discussed the problem and agreed 'that [the Anti-Vivisection] Hospital should not be visited in future in consequence of its work being based on considerations which are not exclusively directed to the welfare of the patients', and that this decision should be kept secret because 'if communicated to the hospital it would at once involve a discussion'.[38]

Their aversion from discussion suggests a lack of confidence in their ability to convince the public that an anti-vivisection hospital was not in the interest of patients. The statistics routinely returned by voluntary hospitals showed the Anti-Vivisection performing at least as well as its peers, with growing patient numbers and a low death rate, but, significantly, neither side showed any interest in comparing the results of cruelty-free versus conventional treatment (for example the Pasteur treatment of rabies versus vapour baths), which indicates that the dispute was not clinical but ideological: were patients best served by animal research to produce better treatments, or by making medicine more compassionate?

Burdett warned the King's Fund that anti-vivisectionists 'are difficult people to tackle, because they always evade the point', though he did not say, and there was no consensus on, what the point actually was.[39] For experimentalists, what mattered most was that vivisection produced results, and they cited the work of Harvey, Pasteur and other medical celebrities to support their case. However, since the outcome of an experiment could not be known in advance, present-day researchers had to rely on anticipated benefits (and what scientist did not anticipate great discoveries?) to justify their continuing use of animals. While some hard-line anti-vivisectionists refused to admit that vivisection had ever led to any useful discovery in human medicine, most accepted that important information had been gained from it in the past, and took the Anti-Vivisection Hospital's pragmatic line that all existing treatments developed through vivisection might still be used, but that no more animals ought to suffer for medicine.

It still took determination, and perhaps a dose of hypocrisy, to take the moral high ground and say of treatments, such as antisera, that were prepared using live animals: 'I had rather die a hundred deaths than blacken my soul by consenting that such deeds should be done for *my* benefit'.[40] If, however, sick anti-vivisectionists lacked the courage of their convictions and accepted antiserum, this did not, as their opponents liked to argue, nullify their position: as the liberal politician Sir Robert Reid (1846–1923) wrote when asked whether he would permit vivisection to save his own family: 'to save them, we might set fire to the nearest cathedral, though we knew it was wrong. Under strong emotion, good men do bad things'.[41]

For anti-vivisectionists, it was, however, better to give into strong emotion than disdain it. The 'forcible suppression of natural feelings of compassion' that vivisection required was 'a ruinous price to pay for knowledge'; ruinous both to the vivisector and to society, since the example of inhumanity was quickly spread. The consequences were particularly hazardous for medical men, since if their sense of compassion were allowed to atrophy they might become as callous to the suffering of patients as they were to that of dogs in the laboratory.[42] Vivisectionists, however, argued that it was they, by forcing themselves to do unpleasant but necessary experiments, who were truly 'on the side of humanity', while anti-vivisectionists selfishly spared their own feelings.[43] The crux of the matter was whether compassion and sympathy were more important attributes for a doctor than detachment and self-control. It was axiomatic that doctors should be caring and compassionate, but self-discipline and fortitude—for example, the 'cool' hand of the surgeon and the persistence of the anatomist during unpleasant dissections—were also essential qualities.[44] Were doctors and their patients forced to choose between sentimentality and science, or was the dichotomy a false one?

Defenders of animal research muddied the waters by pointing out that many anti-vivisectionists seemed indifferent to non-scientific exploitation of animals, such as meat-eating, hunting, and fur-wearing.[45] The British Medical Association, for example, had berated the Welsh landowner Lord Llangattock, a prominent anti-vivisectionist and vice-president of the Hospital of St Francis, for allowing shooting on his estate.[46] Though the BMA was mistaken in this case, there were other anti-vivisectionists who hunted and/or wore fur. In view of these apparent double standards, experimentalists complained that anti-vivisectionists only affected to sympathise with animals, while their true agenda was anti-science and anti-progress.[47] Though this was not entirely true—Oldfield, for example, seems to have been genuinely upset by the suffering he witnessed in slaughterhouses—the New Age movement certainly did aim to reverse the materialistic direction in which science was being taken, and anti-vivisection was an important strategy for doing so. Vivisection too had a socio-political agenda: the controversial nineteenth-century public vivisection displays had been a 'protest on behalf of the independence of science as against interference by clerics and moralists',[48] and formed part of a 'successful struggle by doctors to free themselves and the practice of medicine from older moral and religious concepts and constraints'.[49]

This separation of science from faith and morality proved highly controversial. In *The Perfect Way, or, The Finding of Christ* (1882), Anna Kingsford and Edward Maitland (1824–1897) wrote of a 'falsely so-called' science that concentrated on superficial forms while ignoring the intellectual, moral and spiritual essence of things.[50] To Kingsford, vivisection's brutal materialism was a 'pseudo-scientific inquisition',[51] and vivisectionists, 'black magicians', whose spreading of amorality would 'work havoc' with human souls. The vivisection controversy thus became the new battleground for two divergent views of science, whose proponents had been in conflict since the bodysnatching scandals a century earlier: must science be subservient to the moral codes that governed other human activities, or did its flourishing justify, or even require, setting the old morality aside?[52]

Rallying to the Cause

When the Anti-Vivisection Hospital was incorporated (registered as a business) in 1910, it took a motto that expressed its wider purpose: *Delenda est Crudelitas*—'Cruelty Must Be Destroyed'.[53] In its early years, the hospital sent representatives to anti-vivisection congresses, but the secretary became concerned that this might give the misleading impression they were anti-research. After some of his staff participated in the 1911 BUAV March against vivisection, Robbins went so far as to write to the anti-vivisection journals and popular press to explain that the hospital, which now had twenty-eight beds—twelve medical and sixteen surgical—did not oppose 'legitimate and beneficent forms of laboratory research unconnected with Vivisection such as microscopical investigation of pathological specimens and bacteriological examinations necessary to the interests of medical progress'.[54] In the same year, a new twelve-bedded 'cancer research department', furnished with the latest electric 'light and colour bath', was opened under the direction of Dr Robert Bell (1845–1926).[55] The hospital board was fond of buying such novel pieces of apparatus, probably to show that they were engaging with the latest developments: the out-patient department boasted 'the Buisson Institute for hydrophobia', but no cases of this rare condition ever turned up, and when the inspectors called they found the cupboard in which the Buisson vapour bath was located full of rubbish.[56]

The hospital's death rates were 'exceedingly low'—4.2% compared with the London Hospital's 9.5%—though Robbins prudently told the King's Fund this was not 'a statistic which we would wish to press or labour', presumably because the nature and complexity of the procedures performed in different hospitals would have to have been taken into account.[57] There were a few inquests at which the hospital was censured for poor practice, though it is unclear whether mistakes were more common there or whether its anti-vivisection policy earned it closer scrutiny. Following the death, in 1910, of a woman brought in after an illegal abortion, Dr da Costa was forced to admit in court that they were still not keeping any records, though the jury found the hospital blameless of her death.[58]

Sloppy record keeping and other failings suggest that capable junior staff were hard to obtain, probably because many were unwilling to compromise their careers by linking themselves with anti-vivisection. The resident medical officer was often the only doctor in the hospital, and struggled to cope with its demands: on 4 July 1911, Dr James Hamilton Stuart left the hospital after luncheon, saying he would be back in half an hour, and returned a month later in 'a deplorable condition having been in the East End'. The penniless medic was given breakfast and told to collect his things.[59] Standards among non-medical staff also left something to be desired: the hospital trained its own nurses and was justifiably proud when they passed their examinations, but once qualified they were often left unsupervised, and their work could be 'casual', as was that of the dispenser, who was found to be using an 'extraordinarily large amount of Cocaine'.[60]

Despite such lapses, Battersea's 'working men and women' valued their hospital and organized dances, concerts and boxing tournaments to pay for it, with pro- and anti-vivisectionists working together for 'so worthy a cause in helping the very poor'.[61] In spite of its earlier resolution, the King's Fund did send more visitors, who noted that patients were being treated according to the latest principles, and that no attempt was being made to exclude treatments developed by experiments on animals, but they still refused to make a grant, explaining in an internal memo that: 'The Distribution Committee decided not to give the arguments—which were that the ~~principles~~ restrictions at the Temperance and Homoeopathic hospitals are adopted because they are believed to be in the interests of the patients: those at the Antivivisection Hospital in the interests of the animals'.[62] The antivivisectionists response, had they seen the memo, would presumably have been that the principle of humanity *was* in the interests of patients. What the patients themselves thought would have depended on their views on vivisection, but it seems unlikely the average person would have considered a ban on animal experiments as a restriction, or else why would teaching hospitals have tried to conceal the presence of vivisectionists in their schools?

Whatever one thought of its principles, the Anti-Vivisection Hospital took them seriously: in 1912 an extraordinary meeting was convened to discuss why the resident medical officer, Dr Maurice Beddow Bayly (who by the time of his death in 1962 would be one of Britain's leading medical anti-vivisectionists), had performed a major operation for breast cancer on a terminally ill patient who had shortly afterwards succumbed to surgical shock. There was no suggestion that the patient, who was unconscious throughout, had suffered, and her family did not make a complaint, but Bayly admitted he had not operated for the patient's benefit, but because he had been 'anxious to perform the operation': in other words, he had practiced on a dying woman. This was the kind of 'experiment' that led patients to lose confidence in hospitals, and Bayly, whose conduct had been otherwise excellent, was severely reprimanded for failing to meet the hospital's standards, and left shortly afterwards.[63] In 1926, a nurse was dismissed for 'callous and brutal' treatment of patients, and a visiting doctor was not allowed to demonstrate an Abram's machine because he refused to sign the anti-vivisection pledge; later, a would-be honorary surgeon who tried to water down the pledge by inserting the word 'unnecessary' into the clause about not inflicting pain was not appointed, and a doctor who gave antisera to two patients was asked to resign.[64] Though the hospital was not anti-vaccinationist in the wider sense, vaccines and antisera prepared from live animals were specifically banned under its regulations. The hospital's supporters argued that it was the funds' boycott, rather than hospital's principles, that was harming patients. In 1914, Mary Culme-Seymour (1871–1944) wrote to the King's Fund: 'It can hardly be right that solely on account of the question for or against vivisection, the poor of a populous district should be left without the aid of a Hospital'.[65] Fortunately, the hospital had sufficient income from legacies and donations to remain solvent and even to expand, leaving the King's Fund's obdurate refusal to contribute looking increasingly ineffectual. In 1922, Richard Morris (1859–1956), the MP for Battersea North, called at the Fund's headquarters to ask why they would not assist an institution that was 'doing the work of a general hospital': were they, he asked, 'prejudiced by the views of orthodox medical men on the fund'?[66] That year, the hospital applied to the Fund once again, and was again refused.

It was typical of an anti-vivisection charity that many different kinds of people supported its work.[67] The hospital's chairmen came from varied professional and political backgrounds, and included the libertarian economist Joseph Hiam Levy (1838–1913, chairman until 1905), John Prince Fallowes, Rector of Heene (d. 1941, chairman 1905–1911), the Liberal politician Sir George William Kekewich (1841–1921, chairman 1911–1914), hereditary peer Lord Tenterden (1865–1939, chairman 1914–1926), solicitor Alderman Robert Tweedy Smith (d. 1948, chairman 1926–1927), insurance broker S.C. Turner (chairman 1927–1928) and MP, novelist, theologian and sometime fascist Lord Ernest Hamilton (1858–1939, chairman 1928–1935).

From the beginning, the hospital enjoyed good relations with local trades unions: they thought it was doing 'good work in Battersea' and its committee was 'indebted' to them for helping to raise funds, repaying their support by insisting that hospital contracts went to local firms who paid their men the higher, union-approved wages.[68] There were also strong links with the Progressive Party borough council, which were cemented during the Brown Dog affair, when anti-vivisectionists found common cause with working-class activists. Most of the hospital's funds came, however, from a few wealthy benefactors, notably, in its early years, Lady Portsmouth, whose death in 1906 deprived it of its most generous supporter.[69] Between 1900 and 1914, twenty-four individuals gave over £100 each, including £600 from Viscount Harberton and £500 each from Countess de Noailles and Maynard Geraldine Wolfe. In comparison to this, local street collections and donations from groups with anti-vivisection sympathies, such as the Ancient Order of Druids, brought comparatively little into the hospital's funds, though they helped maintain good relations with the community.[70]

The stereotype of the anti-vivisection donor as a wealthy widow with animals for company was not without foundation: more than two-thirds of the hospital's governors were women, and at one point the hospital planned a charity appeal to ten thousand widows.[71] The preponderance of female supporters led to its becoming known as a women's hospital, despite a lack of female doctors, which led to a costly misunderstanding over the will of a Miss Constance Edith Guerrier of Boulogne, who died in 1926 leaving her entire estate to 'the Women's Hospital, Battersea'. Her executors wrote to the 'Women's Hospital' but as there was no hospital of that name, the letter was delivered to the borough council (now no longer under radical control), which claimed the legacy for its maternity hospital, though this had not existed when the will was written in 1919. The South London Hospital for Women, which was not in the borough of Battersea, then took legal proceedings against the council and won, which alerted the Anti-Vivisection Hospital's Chairman Lord Tenterden, who put in his own claim for the money. Though it emerged during the proceedings that Guerrier, a woman of strong anti-vivisection views, had written to the South London Hospital for Women asking 'if they carried on the Hospital on Anti-Vivisection principles', and had been told they did not, the court ruled that this was the institution to which she had intended to bequeath her fortune, thus depriving the National Anti-Vivisection of £37,000.[72]

In 1922, an inspection of the hospital, which by then was claiming to have fifty-two beds, found only nine beds and two cots occupied: one male ward, one female ward and the children's ward had been closed since the war, along with the private beds in the cancer wing. A new out-patients building, completed in 1915 at a cost of £15,000, remained empty for want of equipment, and the old one was looking run down and antiquated; a faded card on the wall still proclaimed 'RABIES SCARE BUISSON BATH TREATMENT', but the bath had never been used and its gas pipe was not even connected to the mains.[73]

The post-war period was difficult for many voluntary hospitals as there was less money available for charitable giving, and the Anti-Vivisection also suffered from declining interest in its cause. The anti-vivisectionists' ageing demographic was, however, a rich source of legacies, which, typically for a voluntary hospital, were not always listed in its accounts, or even in the minute books. A bequest of £7361 in 1921 wiped off the debt under which the hospital had been labouring since the war, allowing the committee to begin a large building pro-gramme to accommodate the 'ever-increasing applications for admis-sion', and to renovate, redecorate and reequip it 'with every essential modern appliance' until it had been 'made as perfect as possible'.[74] 1929 brought a further windfall of over £38,000 in the form of three large bequests.[75]

Crisis and Compromise

In the 1920s, the hospital was treating over four hundred in-patients and forty thousand out-patients a year, but its reputation was dam-aged by a dispute that began in 1927 when a former honorary surgeon, J.F. Peart, author of 'Foreign bodies in the rectum', made a number of complaints. The hospital's ban on antisera was well known, and Peart claimed that one of his patients had died of tetanus as a result. Furthermore, he told the King's Fund, the hospital 'more than discour-aged' the use of vaccines, two patients had died after straightforward operations owing to the use of infected cat-gut, and 'there were two unqualified persons on the Staff'.[76] In addition, he sent letters, accus-ing 'the women members of the board of management' of interfering with medical affairs, to the *Lancet* and *British Medical Journal*, which printed detrimental reports. In reply, the hospital maintained that Peart had been aware of an arrangement for giving antiserum—by sending patients who asked for it to other hospitals—since 1922.

There are reasons to be sceptical about Peart's allegations. He had been in bad odour with the hospital since 1924, when the 'large fees' he was charging private patients were brought to the attention of its committee.[77] The hospital minutes state he was dismissed after an argument with the matron, though he claimed it was 'because I endeavoured to get the Board of Management to make reforms which the entire qualified Honorary Staff considered necessary'.[78] His claim that the hospital was anti-vaccination is not substantiated by any other sources, the board's interference with clinicians does not seem to have been greater than at other hospitals, and deaths due to infected cat-gut were more likely the result of poor surgical technique than hospital policy. Peart apparently wanted to cause as much damage as possible: he was directed by the King's Fund to stop the hospital receiving ambulance cases, and so decrease its patient numbers and prestige, and this he did by writing to London County Council, which immediately removed the hospital from its ambulance list on the grounds that it 'prohibits the use of serums and discourages the use of vaccines'.[79]

In response, the hospital board told the Council 'that in cases of tetanus and diphtheria and all else medical officers are authorized to take all steps necessary for the preservation of life'.[80] It is not clear how long this compromise had been going on, but it was a damaging admission, as the hospital appeared either unsafe or hypocritical, depending on whether patients had actually been allowed to have antiserum. Some of its supporters, such as the abolitionist BUAV, deplored this concession to expediency, and resignations of governors followed. The County Council took 4 years to reinstate the hospital on the ambulance list, at which time the hospital secretary told the King's Fund that 'In the opinion and practice of the medical staff there is nothing in our constitution or Articles of Association which limits them in treating patients under their care in any way different from that which they practice in other hospitals in which they work. They maintain that sera are not the products of Vivisection'.[81] This removed any pretence that the hospital's practices were different from the norm, which further alienated its core supporters, while making no difference to the King's Fund's boycott.[82]

By the early 1930s, the hospital's financial shortfall had become unmanageable. Eight board members resigned, and the sale of £4300 in investments failed to clear its debts.[83] Late in 1934, the mortgaged hospital received an ultimatum from Barclays Bank, and Lord Ernest made a desperate appeal to the King's Fund: 'My suggestion is that … you, as custodians of the largest British hospital fund, should take the hospital over as the Battersea General Hospital, without the objectionable title of "Antivivisection"'. He added: 'Personally I should welcome relief from my position as Chairman which is burdensome and which occupies much of my time which could be more profitably employed'. The medical and nursing staff were, he wrote, 'second to none' and the building 'modern and up to date in every respect'; it would be a 'calamity' for Battersea if it ceased to operate.[84] In February 1935, Sir Herbert Lush-Wilson (1850–1941) persuaded the Goldsmiths' Company to give the hospital £100 in the expectation that it would soon abandon its anti-vivisection charter, though this was 'not to be mentioned outside the Board Room'. Meanwhile, Lord Ernest continued secretly to negotiate with the King's Fund, which accepted that the hospital was 'geographically' essential.[85]

The Fund agreed to help only if 'Anti-Vivisection' were removed from the hospital's title and charter, which required winding up the company. Realising that this was the Fund's goal, Coleridge tried to press them into assisting immediately on compassionate grounds, but they refused.[86] The board tried to change the hospital's name by a simple vote but this was blocked by the governors, and would in any case not have satisfied the King's Fund, which complained that 'the Hospital's practice was not in accordance with its principles'.[87] Lord Ernest also appealed to various other medical charities, but without success:

The Battersea General Hospital is the only Anti-Vivisection Hospital in the London District and, because of its principles, it is debarred by the prejudice of those in authority from participation in King Edward's Hospital Fund, The Metropolitan Hospitals Sunday Fund, the Charitable bequests of the City Livery Companies and other distributing bodies who generously contribute to the expenses of other London Hospitals. This attitude is unreasonable and unjust because the Battersea Hospital serves a large and very poor district and, by general consent, serves it admirably. … THIS IS PRACTICALLY IN THE NATURE OF AN S.O.S.[88]

Then, in what was either an egregious act of disloyalty or an inept attempt at dissimulation, Lord Ernest told the King's Fund that the governors 'consisted mainly of fanatical women', and the press that '[t]he "anti-vivisection" part of the name of the hospital is a meaningless term pinned to it many years ago, and is a great handicap'.[89] As the *Lancet* noted, this was 'not easily reconciled with the provocative policy for which Battersea has stood in the past'.[90]

Once the situation became publicly known, the King's Fund started to receive complaints about its intransigence. As the Fund was entirely reliant on public money, it seems strange that they so readily ignored public opinion with regard to vivisection, though they may rightly have assumed that committed anti-vivisectionists were unlikely to be among their most generous supporters: they ignored a letter from Coleridge after a check of their records showed he had last made a donation in 1920. They were certainly aware of the problem, as their postbag included letters from exasperated members of the public accusing them of using their influence to 'distort' clinical practice, of failing to honour the memory of Queen Victoria—'whose utter detestation of Vivisection was expressed in perfectly definite terms'[91]—of ignoring the fact that some of their £240,000 annual income must have come from people who were 'suspicious' of vivisection, and of failing to point out when they collected money 'that the Fund was only for the orthodox'.[92] A medical charity had never been less popular: H.R. Maynard, the Fund's long-serving secretary, urged its members to refrain from making any comment about the Anti-Vivisection Hospital, as they 'might not realise the dangers of talking about it outside'.[93] The Mayor of Battersea, a far cry from the outspoken radicals of the early part of the century, refused to allow a public meeting to discuss the crisis, fearing that local people would only use it 'to ventilate grievances'.[94]

The hospital's nurses gave up their holidays to raise money and some local people gave their savings; there was talk of a 'giant' petition asking the Prince of Wales personally to intervene as the Fund's chairman, and Lord Ernest wrote to tell him that 'all Battersea is in a ferment'.[95] The King's Fund sent its representatives to meet the Prince at Ascot races, and after hearing their reassurances, His Royal Highness 'fully realised that it would have been impossible to go back on the previous decisions of the Council, which are based on logical reasons'.[96] There was now only one remaining option, and Lord Ernest told the press that 'If we do not change our name we are dead'.[97] In an internal memo to the King's Fund's distribution committee, its Chairman, Edwin Cooper Perry (1856–1938), wrote: 'Let it die, as it will of inanition if we go on saying nothing'. The ambiguous 'it' may have referred to the bad publicity, but he might just as easily have meant the hospital.[98] Cooper Perry was knighted later that year, for services to charity.

The End of the Experiment

In defeat, Lord Ernest vented his spleen to the press, condemning as 'absolutely outrageous' the 'fanatical bigotry' that had 'killed' the hospital. His outburst made good copy: it was a 'blow for poor people', he said: 'If this hospital was in one of the smart well-dressed districts of London there would be no difficulty at all … But we are situated in a poor district on the wrong side of the river for fashionable sympathy'. When the hospital finally closed at the beginning of June, there were reports of 'sad-eyed' locals watching as ambulances 'whisked' the fifty remaining patients off to other hospitals. Dr John Robert Lee initiated the longwinded process of changing the hospital's charter, informing the high court that in view of the general advance in medical knowledge it was no longer possible to run a hospital on anti-vivisection lines. Mr Justice Lee (no relation) replied that, while he sympathised with the hospital's original aims, he agreed they were no longer practical.[99]

A King's Fund inspection at the end of 1935 found the reopened hospital fit for its purpose and concluded that it had come to grief because of overspending.[100] A number of factors, they felt, had conspired to prevent the hospital paying its debts: declining interest in anti-vivisection in general, disillusionment with the Hospital's equivocation about the antiserum policy, and higher taxation that left the middle classes with less to spare for charitable purposes.[101] In addition, the hospital's policy of paying local people fair wages had kept running costs high, and the anti-vivisection remit made it hard to recruit junior staff and necessitated paying more for those prepared to sign up.[102]

Lord Ernest stepped down as chairman in favour of Sidney Parkes, a London builder and sports promoter, who disconcertingly promised that his strategy to restore the hospital's fortunes would be 'absolutely legal'.[103] Five board members resigned within a year of his appointment, and though he effected 'every possible economy'—reducing the use of X-ray film, requiring attendees at board meetings to pay sixpence for tea and cake, and even accepting a legacy of £100 bequeathed to the (now non-existent) 'Anti-Vivisection Hospital', debts continued to grow.[104] Parkes had personally guaranteed a £6500 bank loan, but there seemed no prospect of repaying it and he resigned in January 1936, 'anxious to be relieved of this responsibility', and having discovered the hard way that 'the Hospital had lost a great deal of income through abandoning anti-vivisection'.[105]

Once its charter had been changed, the hospital's honorary staff were obliged to resign and renew their contracts without the anti-cruelty clause. One who did not do so was Alexander Bowie, the hospital's first chairman, who wrote to the board: 'I remember that we began with one little girl patient. But it soon grew till the number was 100. I appointed the first house physician, a woman, the first nurse and the first medical staff. The hospital was obviously needed, and it has been successful so far as the treatment of patients is concerned...'.[106]

Conclusion

The National Anti-Vivisection Hospital was a thirty-year experiment in putting anti-vivisection principles into practice. It did not directly save a single animal from vivisection, but as a standing protest against cruelty, it became a flagship of the anti-vivisection cause, providing a visible, viable alternative to conventional, 'vivisecting' hospitals and medical schools. For the first time, philanthropists, women's rights activists, socialists, trade unionists, evangelical Christians and other champions of underdogs could help the sick poor without having to compromise their opposition to animal experiments. Like the short-lived but iconic statue of the Brown Dog nearby, the hospital was a rallying point for defenders of humane science, and a provocation to medical orthodoxy. If its efforts to look state-of-the-art by purchasing the latest gadgets looked unconvincing, it at least managed to provide facilities to match those of most conventional district hospitals. Its finances were intentionally kept complex, as like all hospitals dependent on charity it tried to appear needier than it was, but until the 1930s it remained solvent without the help of any of the state-apportioned funds that bolstered the rest of the voluntary system. The fact that there was enough anti-vivisection money to keep even one hospital going was an embarrassment for the medical establishment, and helped convince them that anti-vivisection had great public sympathy and almost unlimited financial support.[107]

In its early years, the hospital made greater efforts to win South Londoners' hearts than their money: 'Give us your help', read a flyer, 'but, above all, give us your sympathy'. Sympathy was an old-fashioned virtue, out of step with the medical profession's push towards dispassionate science, which anti-vivisectionists tried to block by preaching the populist message that knowledge won at the cost of cruelty was not worth the price paid. Experimentalists dismissed such views as sentimentalism or hypocrisy, and tried to shift the focus to what science could achieve rather than the character and methods of those who achieved it. As the twentieth century progressed, the old ideal of the humane, compassionate physician went the way of the honourable soldier and the *noblesse oblige* politician, supplanted by the utilitarian pragmatist who sought the best outcome for the greatest number by any means necessary. By the 1930s, medical scientists had put their case so insistently that it was widely, though perhaps regretfully, accepted among most educated Britons that, as the judge who revoked the hospital's charter put it, medical advances were no longer possible without experiments on living animals. An Anti-Vivisection Hospital could never show that vivisection yielded no benefits: only that a hospital could be run effectively without it, and that there were sufficient conscientious objectors to keep it solvent.

In theory, nothing could have been fairer than the voluntary hospital system: each institution relied on personal philanthropy and received nothing from the public purse; each was governed by a committee answerable to its subscribers, and staffed by consultants who received no salaries. In practice, however, the interposition of the King's Fund and other semi-official charity administrators between donors and recipients led to a system of state-approved patronage. The Fund was what would later be called a QUANGO (Quasi-Autonomous Non-Governmental Organization), answerable in theory to the Crown but in practice to no one outside its own board room (which was the dining room of Marlborough House). Burdett had a talent for getting his own way, but he had no need to persuade a committee that included the Lord Mayor, the Bishop of London, Lords Rothschild and Lister, the Governor of the Bank of England, the President of the Royal Society and the Presidents of the medical and surgical Royal Colleges to refuse support for a hospital that opposed vivisection.[108] The result was that, between 1897 and 1948, the London (later Royal) Homoeopathic Hospital received a total of £70,793, the National Temperance £127,595, and the Anti-Vivisection, nothing,[109] although of the twelve million pounds collected by the fund over that time, thousands must have been given by anti-vivisectionists.

Measured against the criteria used to assess other voluntary hospitals, the Anti-Vivisection held its own. In fact, its work as a hospital was perhaps too successful, since the board pursued their ambitions for growth at the cost of compromising their principles. There was truth in the King's Fund's allegation that by the late 1920s the hospital's management were hypocritically preaching what they sometimes failed to practice, and it was this moral circumlocution, as much as the boycott by the funds, that killed the hospital by alienating its supporters. In 1935, the hospital's board chose to abandon its ideology in order to survive as an institution: a betrayal for its staunch anti-vivisection supporters, but an act of compassion to the sick poor of Battersea, who had been caught up, willingly or otherwise, in an extraordinary and bitterly contested ideological conflict. Even after the board agreed to abandon anti-vivisection, the King's Fund declined to intervene for the sake of the remaining patients, having determined to force the hospital's closure to drive home its point that anti-vivisection was not a viable option.

Shorn of its anti-vivisection trappings, the revived hospital served the people of Battersea until 1972, when it finally succumbed to a National Health Service reorganization. It had been unique in the British hospital system in having been set up and run entirely with anti-vivisection money, but in later years its origins and ethos were purposefully forgotten: the *British Medical Journal's* obituaries of doctors who had worked there in the early days never mentioned the word 'anti-vivisection'. The Hospital's message, that medicine could be practised with compassion, and that some patients preferred this to the cold hand of science, had been so radically unsettling that the establishment wanted it silenced—perhaps the clearest indication of its importance to the anti-vivisection movement.

Notes

1. 'The case of Mabel Florence Jones', *Lancet*, *173* (1909), 124–126; 'Hospital Treatment', *Times*, 31 December 1908, 10 (Hospital Treatment 1908).
2. '"No experiments on patients," but "every kind of operation"', *Nottingham Evening Post*, 31 December 1908, 8 (No experiments on patients," but "every kind of operation 1908).
3. 'Child's death', *Sheffield Evening Telegraph*, 31 December 1908, 2. Many press cuttings on 'human vivisection' are preserved in a scrapbook in Well SA/LIS/E8 (Child's death 1908).
4. Stephen Coleridge, *The Diversion of Hospital Funds: A Controversy between the Hon. Stephen Coleridge and the 'British Medical Journal'* (London: National Anti-Vivisection Society, 1901) (Coleridge 1901).
5. 'Anti-vivisection methods', *The Hospital*, *32* (1902), 244–245.
6. FRCS, 'A guide for the charitable', *Times*, 9 January 1902, 10 (FRCS 1902).
7. Extract from the register of unreported charities, LMA A/KE/245/04; 'Hospital in a lawsuit', *Evening News*, 28 December 1936.
8. Anti-Vivisection Hospital flyer (1903), LMA A/KE/260/001.
9. Letter, G.W.F. Robbins to H.R. Maynard of the King's Fund, 11 November 1907, LMA A/KE/260/001.
10. Hospital board minutes, September 1927, LMA H6/BG/A1/4, p. 137.
11. King's Fund memorandum, *c.* 1912, LMA A/KE/260/001.
12. Extract from visitors' report (1906), LMA A/KE/260/001.
13. On the history of the funds, see F.K. Prochaska, *Philanthropy and the Hospitals of London: The King's Fund, 1897–1990* (Oxford: Clarendon Press, 1992) (Prochaska 1992).

14. Letter, Robbins to Sydney Holland, 18 December 1907; Holland to Robbins, 23 December 1907, LMA A/KE/260/001.
15. 'Ex-Lord Mayor and vicar', *Lancashire Evening Post*, 17 December 1908, 2 (Ex-Lord Mayor and vicar 1908).
16. 'The National Anti-Vivisection Hospital', *Medical Times*, 11 January 1908, 24 (The National Anti-Vivisection Hospital 1908).
17. Prochaska, *Philanthropy and the Hospitals of London*, 53.
18. *Gloucester Citizen*, 24 September 1907, 4; *Times*, 1 October 1907, 10.
19. *Exeter and Plymouth Gazette*, 13 June 1904, 3.
20. Letter, Edmund Hay Currie of the Sunday Fund to Robbins, 15 July 1907, LMA A/KE/260/001.
21. LMA A/KE/245/04.
22. Memorandum, March 1930, quoting a reply from the hospital in 1912, LMA A/KE/260/002.
23. *Exeter and Plymouth Gazette*, 13 June 1904, 3.
24. Letters: Currie to Robbins, 15 July 1907; E.A.G. Pomeroy and Alex Bowie to Currie, 27 August 1907, LMA A/KE/260/001.
25. Peter Mason, *The Brown Dog Affair: the Story of a Monument that Divided the Nation* (London: Two Sevens Publishing, 1997), 55. See also Lansbury, *The Old Brown Dog*; Margo DeMello, *Animals and Society: an Introduction to Human-Animal Studies* (New York: Colombia University Press, 2012), 183–186 and Joe Cain, *The Brown Dog in Battersea Park* (London: Euston Grove Press, 2013) (Mason 1997; Lansbury 2012; Cain 2013).
26. 'The "brown dog" disturbances', *Medical Press and Circular*, 85 (1907), 593.
27. Lansbury, *The Old Brown Dog*, 188; Debbie Tacium, 'A History of Antivivisection from the 1800s to the Present: Part I (mid-1800s to 1914)', *Veterinary Heritage*, 31, (2008), 1–9 (Tacium 2008).
28. Letter, Rose M. George to the Treasurer, King's Fund, 22 May 1930, LMA A/KE/260/002.
29. 'Under the seat', *Chambers's Journal*, 50 (1873), 147–149; A Citizen, 'Vivisection', *Sheffield Daily Telegraph*, 18 May 1899, 7.
30. Buettinger, 'Antivivisection'.
31. Mason, *The Brown Dog Affair*, 8; R.T. Reid, *On Vivisection* (London: National Anti-Vivisection Society, 1902), 6. On anti-vivisection and feminine sentiment see Rob Boddice, 'Equanimity in the Laboratory? The Sentimentalists versus the Sufferers in America *c.*

1900', *History of Emotions—Insights into Research*, October 2013, doi:10.14280/08241.10 (Reid 1902).

32. French, *Antivivisection and Medical Science*.

33. Letter, Pomeroy and Bowie to Currie, 27 August 1907, LMA A/KE/260/001.

34. 'Anti-vivisection meeting in Cheltenham', *Cheltenham Looker-On*, 23 November 1907, 15 (Anti-vivisection meeting in Cheltenham 1907).

35. T. Waldron Bradley. *Torture in Laboratories and Hospitals* (Broadsheet, n.d.), Well SA/RDS A3.

36. Hospital board minutes, 6 August 1908, LMA H6/BG/A1/1; 'The anti-vivisection hospital', *The Hospital*, 25 September 1909, 677.

37. Letters: Savile Crossley to Davis, 21 December 1909, LMA A/KE/260/001; Rose M. George to the Treasurer, King's Fund, 22 May 1930, LMA A/KE/260/002.

38. King's Fund notes, 16 December 1907, LMA A/KE/260/001.

39. Letter, Burdett to Maynard, 28 September 1909, LMA A/KE/260/001.

40. J. Stratton, 'Anti-vivisection and sport', *Medical Press and Circular, 83* (1907), 48 (Stratton 1907).

41. Reid, *On Vivisection*, 5.

42. Edward Berdoe, 'Human vivisections', *Bristol Mercury*, 4 June 1894 (Berdoe 1894).

43. 'Anti-vivisection', *Medical Press and Circular, 69* (1900), 276–277; 'Anti-vivisectionists take note!', *Medical Press and Circular*, (1901), 247–248.

44. Vivisection, like anatomy, was presented as a heroic exercise in self-discipline: see Bates, 'Vivisection, virtue ethics' and Simon Chaplin, 'The heroic anatomist'.

45. This was true of many, though a significant minority were ethical vegetarians: Twigg, *The Vegetarian Movement*, 183–185.

46. F.R.S., 'Lord Llangattock and the anti-vivisection society', *BMJ, 1* (1901), 1239 (F.R.S. 1901).

47. J. Keri Cronin, '"A mute yet eloquent protest": visual culture and anti-vivisection activism in the age of mechanical reproduction', in John Sorenson (ed.), *Critical Animal Studies: Thinking the Unthinkable* (Toronto: Canadian Scholars' Press, 2014), 284–297, on p. 290. For a dissenting view see Anne DeWitt, *Moral Authority, Men of Science and the Victorian Novel* (Cambridge: Cambridge University Press, 2013), 127 (Cronin 2014; DeWitt 2013).

48. Edward Maitland, *Anna Kingsford: Her Life, Letters, Diary and Work* (London: George Redway, 1896), 340 (Maitland 1896).

49. Twigg, *The Vegetarian Movement*, 175.

50. Anna Bonus Kingsford and Edward Maitland, *The Perfect Way; or, the Finding of Christ* (Cambridge: Cambridge University Press, 2011), 31, 184 (Kingsford and Maitland 2011).

51. Mrs Algernon Kingsford, *Violationism: or Sorcery in Science* (Bath: 1887), 1; *Pasteur, His Method and Results* (Hampstead: 1886), 28; MA Cambridge, 'The pseudo-scientific inquisition', *Ethics*, 3 October 1903, quoted in *Zoophilist*, *23* (1903/1904), 155 (Kingsford 1887).

52. Legally there was no cruelty where there was justification: Bates, 'Vivisection, virtue ethics'.

53. Hospital and general purposes committee minutes, 13 September 1910, LMA HO6/BG/A/01/002.

54. Hospital board minutes, 6 December 1911, LMA HO6/BG/A/01/002, p. 151.

55. Special board minutes, 14 June 1911, LMA HO6/BG/A/01/002, pp. 99–102; Hospital board minutes, June 1914, LMA HO6/BG/A/01/002, p. 423. On the light and colour bath see J. Stenson Hooper, 'A new light and colour bath', *Lancet*, *2* (1903), 434 (Hooper 1903).

56. Visitors' report, 1908, LMA A/KE/260/001.

57. Ibid. To this day, politicians and public fail to understand that a high death rate is a consequence of treating the sickest patients.

58. 'Hint to hospital' (newspaper cutting), n.d., LMA A/KE/260/001. It was not unusual at the time not to keep case notes, particularly for out-patients.

59. Hospital board minutes, 5 July 1911, LMA HO6/BG/A/01/002, pp. 110–118.

60. Hospital board minutes, 5 September and 3 October 1907, LMA H6/BG/A1/1.

61. Hospital board minutes, May 1922, LMA H6/BG/A1/3, pp. 344–346; Letter, J. Bench to King's Fund, 26 March 1912, LMA A/KE/260/001.

62. Letter, Maynard to Robbins, 24 May 1912, LMA A/KE/260/001.

63. Emergency committee minutes, 21 August 1912; Hospital board minutes, September 1912, LMA HO6/BG/A/01/002, pp. 215–219, 224.

64. Hospital board minutes, March 1926, April 1926, March 1927, LMA H6/BG/A1/4, pp. 28, 33, 37–38, 106.

65. Letter, M.E. Culme-Seymour to Maynard, 1 June 1914, LMA A/KE/260/001. Mary, one of the two daughters of Admiral Sir Michael Culme-Seymour, was known for the scandal in which it was claimed she had secretly married the young George V.
66. Memorandum, 1922, LMA A/KE/260/001.
67. John Vyvyan, *The Dark Face of Science* (London: Michael Joseph, 1971), 95 (Vyvyan 1971).
68. Hospital board minutes, June 1914, LMA HO6/BG/A/01/002, pp. 415–422.
69. Hospital board minutes, 4 October 1906, LMA H6/BG/A1/1.
70. Hospital and general purposes committee minutes, September 1925, LMA H6/BG/A3/1.
71. There were 247 in total: list of governors, LMA H6/BG/A9/1; Hospital board minutes, June 1933, LMA H6/BG/A1/5, p. 244.
72. Hospital board minutes, May 1928, LMA H6/BG/A1/4, pp. 198–199.
73. Report of Sir Cooper Perry and Sir Frederick Fry following a visit to the Anti-Vivisection Hospital, 22 July 1922, LMA A/KE/260/001.
74. Memorandum, LMA A/KE/245/04.
75. Hospital and general purposes committee minutes, June 1929, LMA H6/BG/A9/1, p. 105. These were estimated residues of estates and may exceed the amounts actually received.
76. Letter, J.F. Peart to Maynard, 28 February 1927, LMA A/KE/260/001.
77. Hospital and general purposes committee minutes, Oct 1924, LMA H6/BG/A3/1.
78. Letter, Peart to Maynard, 28 February 1927, LMA A/KE/260/001.
79. Letter, Cox to the Secretary, Anti-Vivisection Hospital, 12 August 1927, LMA A/KE/260/001.
80. Hospital board minutes, 12 July 1928, LMA A/KE/260/001.
81. Letter, Cox to the Medical Superintendent, 17 March 1932, LMA A/KE/260/001.
82. Letter, Harry W. Woolven to Maynard, 18 July 1932, LMA A/KE/260/001.
83. King's Fund papers, LMA A/KE/245/04; Hospital finance committee minutes, April 1933, LMA A6/BG/A4/1.
84. Letter, Lord Ernest Hamilton to Maynard, 19 December 1934, LMA A/KE/260/002.
85. Hospital board minutes, February 1935, LMA H6/BG/A1/6, pp. 26–27.

86. Letter, Thos H.D. Hogg to the Treasurer, King's Fund, 28 May 1934, LMA A/KE/260/002.
87. Memorandum, 28 April 1932, LMA A/KE/260/002.
88. Letter, Aston Charities Trust to King's Fund, 14 April 1934, LMA A/KE/260/002.
89. Memorandum of meeting between Sir Cooper Perry and Hamilton, 15 January 1935, LMA A/KE/245/04; 'Hospital "killed" by a name', *Morning Post*, 30 May 1935 (Hospital "killed" by a name 1935).
90. 'The name of a hospital', *Lancet*, 8 June 1935.
91. Letter from James Blackwood to Maynard, 9 July 1935, LMA A/KE/245/04.
92. 'Closing a hospital', *Star*, 7 June 1935.
93. Letter, Maynard to Sir Harold Wernher, 19 June 1935, LMA A/KE/245/04.
94. Memorandum of telephone enquiry from A. Bailey, 19 June 1935, LMA A/KE/245/04.
95. 'Battersea General Hospital', *South Western Star*, 23 June 1935; 'Battersea General Hospital fighting for life', *Star*, 15 June 1935, 9; Letter, Sir Godfrey Thomas to Maynard, 11 June 1935, LMA A/KE/245/04.
96. Letter, Wernher to Maynard, 22 June 1935, LMA A/KE/245/04.
97. 'Chance of life for hospital', *Morning Post*, 31 May 1935 (Chance of life for hospital 1935).
98. Memorandum from E. Cooper Perry, 30 May 1935, LMA A/KE/245/04.
99. 'Alteration of a hospital's objects', *BMJ*, 2 (1935), 1055.
100. Distribution committee minutes, December 1935, LMA A/KE/245/04.
101. 'Boycotted Hospital: Battersea ceases to be anti-vivisectionist', *News Chronicle*, 20 November 1935 (Boycotted Hospital 1935).
102. Memorandum of interview with Parker, January 1936, LMA A/KE/245/04; Report of inspection, February 1938, LMA A/KE/512(5).
103. 'Hospital board', June 1935, LMA H6/BG/A1/6, pp. 41–44.
104. LMA A/KE/245/04.
105. Hospital finance committee minutes, December 1937, LMA A6/BG/A4/1, pp. 84–91.
106. 'Hospital in a lawsuit', *Evening News*, 28 December 1936 (Hospital in a lawsuit 1936).

107. Letter, Medical Secretary, BMA to Secretary, MRC, 11 March 1927, Well SA/BMA/C86.
108. Prochaska, *Philanthropy and the Hospitals of London*, 19. The presence of even one woman on the committee might have helped the Anti-Vivisection Hospital.
109. Prochaska, *Philanthropy and the Hospitals of London*, 290.

References

Anti-vivisection meeting in Cheltenham. (1907, November 23). *Cheltenham Looker-On*, 15.
Berdoe, E. (1894, June 4). Human vivisections. *Bristol Mercury*.
Bradley, T. W. (n.d.). *Torture in Laboratories and Hospitals* (Broadsheet).
Boycotted Hospital: Battersea ceases to be anti-vivisectionist. (1935, November 20). *News Chronicle*.
Cain, J. (2013). *The brown dog in Battersea park*. London: Euston Grove Press.
Chance of life for hospital. (1935, May 31). *Morning Post*.
Child's death. (1908, December 31). *Sheffield Evening Telegraph*, 2.
Coleridge, S. (1901). *The diversion of hospital funds: A controversy between the Hon. Stephen Coleridge and the 'British medical journal'*. London: National Anti-Vivisection Society.
Cronin, J. K. (2014). "A mute yet eloquent protest": visual culture and anti-vivisection activism in the age of mechanical reproduction. In J. Sorenson (Ed.), *Critical animal studies: Thinking the unthinkable* (pp. 284–297 on p. 290). Toronto: Canadian Scholars' Press.
DeWitt, A. (2013). *Moral authority, men of science and the victorian novel* (p. 127). Cambridge: Cambridge University Press.
Ex-Lord Mayor and vicar. (1908, December 17). *Lancashire Evening Post*, 2.
F.R.S. (1901). Lord llangattock and the anti-vivisection society. *BMJ, 1*, 1239.
FRCS. (1902, January 9). A guide for the charitable. *Times*, 10.
Hospital Treatment. (1908, December 31). *Times*, 10.
Hospital "killed" by a name. (1935, May 30). *Morning Post*.
Hospital in a lawsuit. (1936, December 28). *Evening News*.
Kingsford, A. (1887). *Violationism: Or sorcery in science* (Vol. 1). Bath.
Kingsford, A. B., & Maitland, E. (2011). *The perfect way; or, the finding of Christ*. Cambridge: Cambridge University Press.

Lansbury. (2012). *The old brown Dog*, margo demello, *animals and society: An introduction to human–animal studies* (pp. 183–186). New York: Colombia University Press.

Maitland, E. (1896). *Anna kingsford: Her life, letters, diary and work*. London: George Redway.

Mason, P. (1997). *The brown dog affair: The story of a monument that divided the nation*. London: Two Sevens Publishing.

"No experiments on patients," but "every kind of operation". (1908, December 31). *Nottingham Evening Post*, 8.

Prochaska, F. K. (1992). *Philanthropy and the hospitals of London: The king's fund, 1897–1990*. Oxford: Clarendon Press.

Reid, R. T. (1902). *On Vivisection* (p. 6). London: National Anti-Vivisection Society.

Stenson H. J. (1903). A new light and colour bath. *Lancet, 2,* 434.

Stratton, J. (1907). Anti-vivisection and sport. *Medical Press and Circular, 83,* 48.

Tacium, D. (2008). A history of antivivisection from the 1800s to the present: Part I (mid-1800s to 1914). *Veterinary Heritage, 31,* 1–9.

The National Anti-Vivisection Hospital. (1908, January 11). *Medical Times,* 24.

Vyvyan, J. (1971). *The dark face of science*. London: Michael Joseph.

6

The Research Defence Society: Mobilizing the Medical Profession for Materialist Science in the Early-Twentieth Century

At the end of the nineteenth century, medical experimenters were on the defensive. Although the 1876 Act had done nothing to reduce animal use, and Britain's physiologists were confident they could ensure it was 'harmlessly administered' by putting 'effectual pressure upon officials...',[1] it was still more restrictive than the laws of any other European nation, and more intrusive than any of them wanted: one American observer thought it 'significantly handicapped the teaching (if not the practice) of British physiology'.[2] Anti-visectionists naturally claimed the Act did not go far enough, and used initiatives such as anti-vivisection hospitals and pamphlet wars to raise doubts over whether experiments on animals were necessary, and fears that they would inevitably lead on to human experimentation.

The nineteenth-century medical profession had been able to extract some positive publicity from the bitter dispute over vivisection by presenting it as a difficult and demanding task that required great fortitude and commitment to science. Flouting conventional sensibilities had, after all, served medics well in the recent past, when body snatching scandals had introduced the trope of the 'mad' scientist, a heroic figure who transgressed moral boundaries not for personal gain but to win

© The Author(s) 2017
A.W.H. Bates, *Anti-Vivisection and the Profession of Medicine in Britain*,
The Palgrave Macmillan Animal Ethics Series, DOI 10.1057/978-1-137-55697-4_6

valuable knowledge for the benefit of mankind. The public were willing to be persuaded that the anatomist's noble ends might justify the base and sometimes illegal means used to obtain corpses, though their response was strangely ambivalent: angry crowds called for body-snatchers to be hanged, while anatomy shows became so popular that the subject 'turned to gold'.[3] The furore certainly made it clear to everyone that, after the Anatomy Act, doctors enjoyed privileged access to a body (of course) of knowledge that was denied to laymen, who had neither the opportunity nor the stamina to attend the dissections necessary to obtain it.

In the late-nineteenth century, vivisection assumed a similar function, and vivisectionists, like anatomists before them, enjoyed an ambiguous reputation as both perpetrators of atrocities and pioneers of science. In the twentieth century, however, when vivisection was set up as one of the pillars of experimental medicine, it could hardly remain the prerogative of a few eccentric, taboo-breaking innovators. It had to be normalized within a medical culture that based its intellectual and moral authority on the intimate relationship between professional knowledge and laboratory experimentation.

An obvious obstacle to this was the steady growth of anti-vivisection organizations, which kept up the pressure on experimenters by publically questioning their judgement in using animals to study human disease.[4] According to Tansey, it was the antis persistence that led to the setting up of another Royal Commission in 1906.[5] By this time, the use of laboratory animals was far greater than when the first commission had reported 30 years earlier: the nationally-funded Imperial Cancer Research Fund (ICRF) alone, founded in 1902, was already performing 8,600 experiments a year.[6] Although these were all carried out in accordance with the 1876 Act, large increases of this kind would probably not have commanded public support, and so vivisectionists tended not to publicise them. A new anti-vivisection strategy arose of publishing details of how many experiments were being performed in the hope of shaming vivisectionists into stopping: the *Anti-Vivisection Review* called this publicity 'The Light dreaded by all Vivisectors'.[7]

The second Royal Commission offered an important opportunity for experimenters to launch a 'counter-attack' in the propaganda war. The medical community initially coordinated their response through the Association for the Advancement of Medical Research (AAMR), a working group that had been formed in the wake of the Cruelty to Animals Act, to advise the Home Office on granting licenses. Concerned that the Commission planned to tighten up the regulations, a number of leading physiologists decided to found a new group, the Research Defence Society (RDS), whose inaugural meeting on 27 January 1908 in London's Harley Street saw Lord Cromer elected as President and Stephen Paget (1855–1926) as Secretary. It consisted of a small but distinguished all-male group of physiologists, most of whom had links with UCL, which, thanks to the Brown Dog riots, was the best-known centre for animal experimentation in the country. Among those present at the first meeting were Professor Cushny,[8] Sir Victor Horsley (1857–1916), Dr Charles Edward Beevor (1854–1908), Dr Leonard Hill (1866–1952), Lord Justice Fletcher Moulton (1844–1921) and Sydney Holland. The RDS received a start-up grant of £185 jointly from the Physiological Society (another pro-vivisection group, founded in 1876 'for mutual benefit and protection') and Professor Ernest Starling's (1866–1927) UCL working group, which was busy preparing evidence for the Commission.[9]

The Earl of Cromer, a diplomat and banker, had very little to do with the day to day running of the Society. That task fell to Paget, who was also Secretary of the AAMR, many of whose members joined the fledgling RDS, the two organizations eventually amalgamating in 1917.[10] Paget, whose role was somewhat akin to that of Stephen Coleridge in the anti-vivisection movement, was not himself a vivisectionist and had no connection with UCL. He was medically qualified, but had given up practice altogether in 1910 to devote his professional energies to justifying vivisection, writing frequent articles and letters, and lecturing widely on the topic, despite his worsening health.[11]

Most of the other RDS members were active researchers. Sir Victor Horsley, FRS, who had trained at UCL before becoming Professor of Surgery there, was a man of strong social principles—a temperance reformer and supporter of women's suffrage—and was particularly noted for his kindness to patients. In writings such as 'The Morality of Vivisection', he argued that excessive feeling for animals was displacing proper concern for humans. His own concern for humanity led him to vivisect over 3000 animals (his work on gunshot wounds in dogs involved shooting them in the head), a record hotly criticised in the correspondence columns of *The Times*. He had every reason to resent antivivisectionists, having been professor at the notorious Brown Institute in London's Vauxhall (a post previously held by Burdon-Sanderson, author of the vivisectors' handbook), an institution the BUAV and others had strenuously, though unsuccessfully, campaigned to close.[12] It was Horsley who had helped persuade his colleague Bayliss to bring the so-called 'Brown Dog' libel action against Stephen Coleridge in 1903 for reading out an extract from *The Shambles of Science* at a public meeting—UCL men were particularly riled by the chapter entitled 'Fun', in which Lind af Hageby described their 'jokes and laughter' during experiments.

Despite being a dog owner himself (a photograph shows him holding an understandably nervous looking Jack Russell), Horsley was spokesman for the Society for the Prevention of Hydrophobia, and a leading advocate of compulsory dog muzzling—the Society's answer to Britain's periodic outbreaks of 'hydrophobia'. These epidemics took the form of a rash of press reports about 'mad' dogs, at least some of which were probably suffering from common canine diseases such as distemper, or simply misbehaving, rather than carrying the dreaded disease.[13] How much genuine rabies there was in early-twentieth century Britain remains an unanswered question, but scare tactics from the Society for the Prevention of Hydrophobia, such as recommending that all dogs had to be kept muzzled in public in order to prevent a national rabies epidemic, were certainly expedient for vivisectionists, for whom unwelcome and unwanted strays were an easy source of low cost experimental subjects.[14]

Another two UCL men at the inaugural RDS meeting were Leonard Erskine Hill, a physiologist working on decompression sickness, and who held a high view of research ('the path which saves the millions when found'),[15] and Beevor, an associate of Horsley.[16] An important non-medical supporter was Moulton, a barrister who gave evidence to the Royal Commission; he 'preferred not to be a member', in case this was thought to conflict with his legal work, but after he became an Appeal Court judge in 1912, and the first chairman of the Medical Research Council (MRC) the following year, he used his influence to help the RDS 'again and again'.[17]

The second Royal Commission's report was a foregone conclusion given its skewed composition: naturally enough, there were several vivisectors on the panel, but the BUAV's attempt to have Walter Hadwen appointed was rejected on the grounds that as an anti-vivisectionist, he would be biased. Between 1906 and 1912 the commission sifted a large amount of evidence, before unsurprisingly endorsing the status quo. Their major recommendation was the usual one for a public enquiry desirous of seeming to act while in fact doing nothing: they set up a committee to take over the AAMR's role of advising on licensing, which in the BUAV's view made no difference. While the Commission was still pondering the evidence, vivisection had received a boost from the 1911 National Insurance Act, which set aside state (i.e., taxpayers') money to fund medical research.[18] Two years later, the Medical Research Committee and Advisory Council (later the MRC) was formed to control how the money was allocated.

The Early Years

Membership of the RDS was by invitation only and within a month of its foundation almost three hundred experimenters had joined, only six of those approached having declined. The annual subscription was 5 s and life membership £10: since one would have had to live another 40 years for life membership to work out cheaper, the Society apparently had young experimenters in its sights, but as there are no extant membership lists, we cannot be sure who joined. The secrecy was not because members would have faced any direct threat—violence against vivisectionists was unknown until the 1970s—but because being known as a vivisectionist might have deterred some donors, and patients. After only a few months there were over a thousand members, including eighty-four women, plus many hundreds of associates, mostly medical students, who paid half a crown.[19]

The RDS quickly acquired a list of distinguished vice-presidents to rival those of the anti-vivisection societies: these included Elizabeth Garrett Anderson, Lord Curzon, the Duke of Devonshire, Edward Elgar, M.R. James, Rudyard Kipling, Ray Lankester, William Osler, and Henry Wellcome.[20] To the dismay of the anti-vivisection camp, sixteen Anglican bishops signed up, perhaps desirous of demonstrating, as the Bishop of Edinburgh had stressed at a church congress 20 years earlier, that the Church was not 'anti-science'.[21] RDS meetings were, and remained, ticket only, despite being described as 'public', but interested laypeople could, and did, attend; at a meeting of the Kensington branch, more than half the audience were women.[22]

The initial strategy was to hold educational meetings and provide speakers for public debates. The society kept itself abreast of the times and locations of anti-vivisection meetings, and accepted 'challenges' to send along a speaker. These were competitive debates, with a vote taken at the end, and the RDS minutes record that they 'won' in Winchester and Bow, but lost in Bath and Oxford. However engaging these sessions were to those present, the small numbers voting—usually less than a hundred—suggests they did not have a very wide appeal. Recognising that competitive debating was unlikely to achieve its objects, the RDS decided not to accept any further invitations from anti-vivisection societies, though they continued to provide public speakers into the 1920s.[23]

Both the RDS and their opponents were, however, willing to continue the argument at a less intellectual level. In January 1909, the committee was shown a circular, signed 'M. Cowan', outlining 'a plan of prayer for the sudden death of this or that person making experiments on animals'. This echo of Anna Kingsford's notorious campaign of psychic assassination, which she claimed had brought about the deaths of Claude Bernard and Paul Bert, was the sort of anti-vivisection excess that the RDS quite reasonably thought would help their cause. Unfortunately, their own tactics were scarcely more subtle; they accused Lord Llangattock of being 'addicted' to stag hunting, a libel that Dr Morgan Jones for the RDS was foolish enough to try to defend, leaving the Society with no choice but to publish an apology in order to avoid being sued, though Paget churlishly limited it to 'a purely formal expression of regret'.[24]

The task of the RDS was made easier by the anti-vivisectionists' inherent disunity. A major schism had occurred in 1898 when the National Anti-Vivisection Society passed the following resolution by 29 votes to 23:

The Council affirms that, while the demand for the total abolition of vivisection will ever remain the ultimate object of the National Anti-Vivisection Society, the Society is not thereby precluded from making efforts in Parliament for lesser measures, having for its object the saving of animals from scientific torture.[25]

As a result, Frances Power Cobbe had set up the British Union for the Abolition of Vivisection, which demanded a total and immediate ban on all animal experiments, while the NAVS, under Stephen Coleridge's leadership, was prepared to accept lesser measures in the interim. The choice between principled total abolition and pragmatic gradualism was to prove an enduring source of tension for the movement.

In 1909, two rival international anti-vivisection congresses were held in England, one organised by Lizzy Lind af Hageby's gradualist Animal Defence and Anti-Vivisection Society, and the other by the World League Against Vivisection, which was committed to total abolition. While the latter's position was perhaps logically more coherent (if vivisection were morally wrong, there could be no excuse for tolerating it), it seemed to the pragmatists that those demanding 'all or nothing' were liable to end up with nothing. For the World League, however, total abolition seemed achievable, not least because they had the support of several prominent Labour MPs, whose party was a rising force in parliament.[26]

When engaged in debate, the RDS favoured the time-honoured utilitarian position that vivisection saves human lives, which they backed up by providing facts about the important discoveries that Pasteur and other famous medical scientists had made through animal research. It could be difficult for their some of their speakers to appreciate why the audience remained unconvinced by such powerful evidence. In 1910, the distinguished astronomer Sir David Gill (1843–1914), speaking for the Society, was involved in the following exchange:

Gill: 'Let any mother whose child is suffering from that dangerous disease diphtheria be asked how many dogs' lives she would give for the life of her child'

A lady in the audience: 'None'

Gill: 'God help humanity! (Applause and a few hisses)
                   One lady says she would not give one'
Lady: 'Not one'.[27]

The RDS undertook direct action against anti-vivisectionists with
more success. They asked railway companies not to display anti-
vivisection posters on their stations, with the result that the District
Railway stopped accepting them, while the canny Great Western continued to take the anti-vivisectionists' money but placed the posters in
obscure positions.[28] The RDS also persuaded railway bookstalls not
to sell anti-vivisection pamphlets, asked the publisher Sir Frederick
Macmillan to drop anti-vivisection publications from his list, convinced
the Postmaster General to ban anti-vivisection advertisements in post
offices, and induced the organisers of Cruft's dog show to stop hosting
a BUAV stall there.[29] They even planned to break up a 'monster proces-
sion' in the anti-vivisectionists' home territory of South London, until
they discovered there would be 'a large contingent of Battersea roughs,
to protect the banners from medical students…'.[30]

A common means of promoting the anti-vivisection message was to
rent display space in a shop window and hand out leaflets outside. The
RDS hired men with sandwich boards to March up and down in front
of the ADAVS's 'anti-vivisection shop' in London's bustling Piccadilly
and hand out leaflets of their own, a tactic that was successful insofar
as the shop closed after a year, though it probably did little to spread
a positive message about research. Passers-by found these aggressive
pamphleteers, who came to blows on one occasion, a nuisance, and
Westminster Council complained about the resulting litter of leaflets that beleaguered pedestrians threw away as soon as they could: a
salutary reminder that most people had no interest in the vivisection
question one way or the other.[31] In 1910, the BUAV opened shops in
Wimbledon, Newcastle, Southport, Liverpool and Worcester, and the
minutes of the RDS note that '[e]ach of these shops had been duly
besieged with the leaflets of the Research Defence Society, and had been
closed'. The contents of their leaflets was as disingenuous as those of
the antis: when George Robbins of the Anti-Vivisection Hospital wrote

to Paget complaining about a pamphlet entitled 'Fighting the Invisible', Paget admitted that it was untrue to state that animals were always killed immediately after experimentation, because that 'would defeat the objects of enquiry', but said the RDS was not responsible for the contents of the pamphlets they distributed.[32]

The fuss stirred up by the second Royal Commission and the ensuing pamphlet wars died down with the outbreak of the First World War. The public had other concerns, and newspapers had little space to report domestic spats. Civilian medical research and education were stepped down as universities focussed on training extra doctors quickly, and many anti-vivisection groups suspended their activities,[33] though their work was not done, as military experiments replaced civil ones. As the conflict escalated into an industrial and technological race to manufacture weapons faster and in greater numbers than the enemy, science was enlisted to come up with new ones: in 1915, the Ministry of Munitions acquired six thousand acres in Wiltshire that became the War Department Experimental Ground, better known as Porton Down, and commandeered a nearby farm to breed animals for the chemical and other weapons tests that were set to take place there.[34]

Wartime statistics were presented in such a way as to obscure the large increase in the number of animals being used; for example, in 1916, the *Lancet* noted that the latest Home Office return showed 2771 experiments had been performed the previous year, 'other than those of the nature of simple inoculations, hypodermic injections, or similar proceedings…'.[35] While this sounded reasonable enough, these 'simple' injections included inoculations with tuberculosis, anthrax, rabies and bubonic plague, and there were many thousands of them.[36] National newspaper proprietors were bound to keep the government's secrets, and were provided with positive stories about how research on animals was leading to major advances in the fight against disease amongst the troops. The anti-vivisection lobby unfortunately chose to counter what they saw as the government's increasingly authoritarian attitude to medicine by campaigning against compulsory vaccination, with the result that they were denounced in the House of Commons as 'unpatriotic' for trying to stop soldiers from receiving life-saving inoculations.[37] Since this could be construed as interfering with the war effort, a crime under the Defence of the Realm Acts, the anti-vivisection movement found itself effectively boycotted by the press.[38]

Inter-War Politics

After the war, the RDS claimed that experiments on animals had saved the lives of many wounded men by helping surgeons to understand 'the violences offered in modern warfare to the human body…'[39], and they branded the anti-vivisection movement 'an enemy of the people'.[40] The Society resumed its course of blocking anti-vivisection initiatives whenever it could while eschewing open debate, but Paget, whose health was deteriorating, became concerned that, although public opinion was turning in their favour, it would be impossible for them to overcome opponents who had many times more money to spend than they did (in 1921, the RDS's annual income was under £750—less than one anti-vivisection society was spending on stationery).[41] 'We cannot follow them everywhere', he wrote despairingly, painfully aware that many local RDS branches were 'half-dead'.[42]

Fortunately for the RDS, the war had sapped support for anti-vivisection also. Appalling though the suffering inflicted upon animals during the conflict had been, it was lost in the shadow of the human tragedy. It was also believed that animal experimentation carried out in secret had helped the war effort. The BUAV made the best of it, asking donors to recall that animals too had served their country by performing military service, but in the post war years many people were poorer, taxation had increased, and there was little money to spare for animal charities. Most of the anti-vivisection movement's charismatic founders were now dead or aged, and the societies they had founded, like all protestants, had proved fissiparous, and, rather than pooling their resources, tended pursue separate agendas. Consequently, most of the anti-vivisection work undertaken in the inter-war period was uncoordinated, and met with scant success. High-profile researchers such as the UCL physiologists were still subjected to attacks in print, which was how Lizzy Lind af Hageby and Leisa Schartau had begun their campaign at the turn of the century, but perhaps the only new tactic of note was a 'van campaign' to transport posters and pamphlets to far-flung parts of the country and collect signatures for petitions.[43]

Within the medical profession, the growing reliance on animal experiments seemed to have gained unstoppable momentum: animal research had come to be regarded as the gold standard and so scientists who wanted their work to be taken seriously by their peers more or less had to use animals. Of the Nobel prizes in physiology or medicine awarded in the hundred years from 1901, only thirteen did not involve research on non-human vertebrates. This pattern was reflected by lesser prize- and grant-giving bodies, and so animal research became linked in the public mind with reports of acclaimed new discoveries, while the details of what actually went on in the privacy of the laboratory were usually glossed over.[44]

There were some attempts by the anti-vivisection lobby to have Porton Down closed, but they came to nothing as the government was keen to continue its research in anticipation of another conflict. Curiosity about this secret establishment led to speculation in the press and questions in parliament, where the answers given only hinted at what might still have been going on there. In May 1923, the Under Secretary for War admitted in the House of Commons that over 700 animals, including 23 monkeys, had been used in 'gas poisoning' experiments the previous year, and the number of animals acknowledged to have been used in such testing continued to rise until 1925, when the Geneva protocol banned chemical warfare.[45]

Ironically, peace initiatives such as the League of Nations, formed in 1920 to prevent future conflicts, tended to lead to more experiments on animals, since improvements in 'modern medicine' were hyped as a means to alleviate the social hardships that led to war.[46] The post-war conflict in which Europe was now engaged was a war against poverty and disease, and it was being fought with modern, scientific weapons developed in the clinic and laboratory. Pro-vivisection literature made increasing use of this military imagery, with the RDS reporting what had been learned from the 'sacrifice' of laboratory animals in a series of pamphlets entitled *The Fight Against Disease*. Like any war, it needed large sums of money to be raised at a national level: to this end, the British Empire Cancer Campaign was founded in 1923, to supplement the work of the ICRF and MRC, who initially considered it a rival.

The idea that science held the keys to health, prosperity and peace naturally fostered a positive attitude on the part of the public towards laboratory research: at the Efficiency Exhibition in Olympia there were long queues at the Middlesex Hospital stall as people waited to look down the microscopes.[47] The British experienced nothing like the reaction to laboratory-based work that had been seen in Germany, where a so-called 'crisis of medicine' occurred as many ordinary people lost their trust in 'mechanistic' medicine and turned instead to heterodox practitioners of alternative medicine, whose treatments seemed more natural and whose methods appeared more patient-centred.[48] In Germany, vivisection became a symbol of the failure of scientific medicine to respect the ideals of the traditional healer, and the public responded enthusiastically when the rising National Socialist Party adopted an anti-vivisection policy.

In Britain, those parts of society that were most outspoken with regard to vivisection failed to hold the government to account when they were able to, presumably because they felt there were more important issues to vote on. Anyone who had predicted that the government would be forced to ban vivisection once women and the non-landed classes were enfranchised turned out to be badly mistaken.[49] The first Labour government, formed in 1924, included no fewer than four cabinet ministers—Ramsay Macdonald (1866–1937), Philip Snowden (1864–1937), Arthur Henderson (1863–1935) and J.R. Clynes (1869–1949)—who had been pre-war supporters of the World League Against Vivisection, but all of them failed to put their purported principles into practice now that they were able to do so. Clynes confessed his hypocrisy in the most elegant of phrases, writing that in the matter of vivisection he found it impossible 'to harmonise his public duties with his private opinion'.[50]

With support for anti-vivisection groups on the wane, the RDS turned its attention to ensuring that the income derived from legacies as wealthy anti-vivisectionists died off was minimised. In a landmark legal case in 1928, the RDS appealed against a legacy of £200,000 being used to set up a trust to fund the Beaumont Animals Benevolent Society.[51] The bequest was as bizarre as it was generous, but the Court of Appeal ruled that its purpose—to create a sanctuary where all kinds of animals could live undisturbed by humans—was not charitable, since the Court reckoned that this would provide no benefit to the public.[52] Although this so-called Grove-Grady case was not related to anti-vivisection, the judgement did establish a valuable principle, from the RDS's perspective, that money given *for the benefit of animals alone* could not be considered charity. Had the court decided otherwise, wrote its secretary, it 'would have given us endless trouble',[53] as it was, the decision paved the way for a future ruling that anti-vivisection too could not be a charitable cause.

For veterinary hospitals where research on animals took place, Grove-Grady meant that charity must be given for the benefit of the hospital rather than the animals. In 1931, the RDS took legal action against the BUAV on behalf of the Tail Waggers Club, a fundraising scheme for the Royal Veterinary School that the BUAV had tried to block on the grounds that it subsidised animal experimentation. The RDS's action was successful, winning for Captain Hobbs, the only human member of the Tail Waggers, £500 in damages.[54] The following year, the School found themselves less grateful for principle that charity could not be given solely to help animals when the RDS threatened legal action to prevent them accepting £25,000 from an anti-vivisectionist, though the Society relented after School promised 'that none of the sum be devoted to anti-vivisection propaganda'.[55]

The RDS did not receive government assistance, remaining essentially a private lobby group for vivisectionists. New license holders received a letter inviting them to join, but neither the identities of those who did, nor the total number of members, was made public, though there must have been far fewer than even the smallest anti-vivisection society, and the RDS's annual income was less than a thousand pounds. Considering its modest budget, the influence it was able to exert was impressive. In 1934, the honorary secretary, G.P. Crowd, summarised its principal achievements: the defeat of the Dogs Protection Bill, the defence of University College in the dog stealing case, changing the attitude of the RSPCA to research, blocking the Grove-Grady bequest, and protecting research at the Royal Veterinary School.[56] A closer examination of the first of these, the long and frustrated progress of the Dogs Protection Bill in its various forms, shows how the RDS was able to influence parliament and collaborate effectively with the BMA and other pro-vivisection groups.

Stopping the Dogs Protection Bill

The Dogs Protection Bill was conceived as a means to break down the vivisectionists' defences at their weakest point, necessarily forgoing the support of total abolitionists (though the BUAV backed it) in the hope of winning a tactical victory on behalf of the species most adept at appealing to human sympathy. The popularity of dogs as domestic companions as well as their usefulness as working animals made their use as experimental subjects seem particularly objectionable; as Lord Dowding put it, 'the dog has no aim in life other than to love and serve humanity', and the relationship between dogs and humans was often assumed, with good reason, to be a special one.[57] Dogs had had their own anti-cruelty lobby since 1891, when the National Canine Defence League was founded by the breeders of show dogs, to protect dogs from all kinds of cruelty, from vivisection to muzzling.

The advocates of dog-specific anti-vivisection legislation left themselves open to the criticism that they were acting on an irrational, sentimental bias towards a favourite pet, and indeed the positions of both sides in the dogfight were not far removed from hypocrisy. Experimenters pretended there was nothing special about dogs, but then admitted they preferred them because they were particularly cooperative and biddable, even under torture, an admission condemned by anti-vivisectionists as the heartless betrayal of a friend and helper. Dog lovers, however, had little to be proud of; one reason that dogs were such a popular laboratory animal being that they were readily obtainable: Britain was home to a large underclass of strays and mongrels which, though legally protected from vivisection by the 1906 Dogs Act, were, in reality, like pauper cadavers a century before, worthless to all except experimenters, who could easily acquire them for money.

In 1906, when the BUAV sought support for a dogs protection bill (not an entirely novel idea, since a 'Dog Protection Bill' had been contemplated as long ago as the 1840s to prevent the theft of 'fancy dogs' for export),[58] the public, understandably averse from the idea of anyone experimenting on what they saw as pets, responded enthusiastically, and the BUAV was able to present the Home Secretary with a petition weighing a quarter of a ton, nine miles long, and with over 400,000 signatures.[59] Although the size and weight of this monster petition were obviously inflated for dramatic effect, it still stands as one of the largest written petitions in English history.

The BUAV had appealed to the public because a similar bill introduced in the House of Commons the previous year had failed to progress.[60] For those unversed in the labyrinthine complexities of the British parliamentary system, bills receive a nominal 'first reading'— in which the title is read out and the bill is ordered to be printed— followed by a second reading when they are debated and voted on. If they are passed, there follows a committee stage at which amendments are made and voted on, before the bill may proceed, time permitting, to a third reading and another vote. Once a bill has been passed by the Commons, it must then undergo a similar process in the House of Lords, and only when this is finally complete is the bill passed for royal assent, at which point it becomes law. Few private member's bills, i.e., those not sponsored by the government, ever become law, unless the government aids their passage.

The medical profession, through the BMA, responded to the huge petition by releasing a strongly worded manifesto opposing the bill,[61] which suffered the fate of most unsupported bills, reaching its second reading but then running out of parliamentary time. It was reintroduced in 1907 by the radical liberal barrister Ellis Griffith (1860–1926), but met with the same outcome. Sir Frederick (later Lord) Banbury (1850–1936) tried again in 1908, to loud cheers from the backbenches, but despite the support of the fifty or so MPs who were members of the BUAV-sponsored Parliamentary Anti-Vivisection Committee, the bill was blocked by members representing medical and university interests. In 1911, after the BUAV's parliamentary question on the legality of selling dogs for experimentation received an 'evasive and unsatisfactory' answer from the government,[62] Banbury vowed to reintroduce the dogs bill in every session until it was passed. The bill passed its second reading in 1913, but was 'wrecked' at the committee stage, when it was decided, by thirteen votes to twelve, to amend it to allow experiments performed under anaesthesia,[63] thereby, quipped one newspaper, ensuring that the bill was 'painlessly killed'.[64] Perhaps the most quotable contribution in this debate came from the G.G. Greenwood (1850–1928), a supporter of the bill, who, when asked by the committee whether he would vivisect a dog to save his own child, replied: 'to save my child, I should, very possibly, be prepared to vivisect the honourable Member who asked [that question], But that would hardly be accepted as proof that I was morally right in so doing'.[65]

The Bill returned again in 1914, to the dismay of the BMA, and 350 physicians and surgeons wrote to *The Times* in protest.[66] This was sufficient to mobilise enough MPs to block it as, apart from the parliamentary anti-vivisection group, most had no strong opinions on the fate of stray dogs. Accused of being motivated by sentimentality, the anti-vivisection MP Colonel E.S. Sladen said he was 'proud' of being sentimental about dogs, and would welcome 'the advent of sentiment into the house of commons'.[67]

The tenacious Banbury ('a grim old Tory, but he has a very soft heart for a dog') tried again in 1919 and the bill 'slipped through' its second reading in spite of the efforts of the Commons medical committee, who were 'caught napping', and whose best argument seemed to be that the bill was against the national interest, since research on dogs during the war had led to the development of gas masks.[68] The bill got through to the committee stage but was amended to make it, in the words of the RDS, 'absurd' and then defeated on a three line whip (i.e., the government compelled its MPs to vote against it on pain of expulsion from the party).[69] It did not help that an 'epidemic' of 'rabies' broke out while the bill was before the house, a coincidence that looked to one newspaper like 'a political dodge' to deprive stray dogs of public sympathy.[70] *The Times*, which opposed the Bill, took the opportunity to remind readers that the cure for 'this most awful disease' had been discovered by Pasteur's experiments on animals.[71]

The medical lobby continued to present vivisection as both essential and innocuous. A deputation from the Royal Society of Medicine told the Home Office that 'in the absence of infection the wounds [of vivisected dogs] were not painful', but it is inconceivable that anyone with medical experience actually believed this; the point was well made by anti-vivisectionists that scientists were never willing to have these 'painless little operations' performed on themselves.[72] The deputation also appealed to the national interest by claiming that 'success in war or in industry was bound up in experimental research', which gave the government a political and financial excuse for permitting as many experiments as possible, while making anti-vivisection seem to go against the national interest.

The established church, or at least those in authority within it, sided with the RDS. The (Protestant) Archbishop of Dublin, who was president of the RDS's Dublin branch, complained that the bill would 'retard the advance of medical knowledge and hinder the work of surgeons for the benefit of suffering mankind'. He told an RDS meeting that, compared with human needs, the interests of animals were morally negligible: 'Man has a dignity of his own which he does not share with the lower creatures. He is an "end in himself", as the philosophers say: you cannot say that of any other animal'.[73] Pro-vivisection bishops earned the disapproval of some members of their flocks, but their lordships held firm in their views: the Church Anti-Vivisection League told Bishop Frodsham of Queensland that his involvement with the RDS, 'having nothing to do with his sacred office, is a scandal and cause of offence to manifold members of his flock', but the bishop replied that vivisection was a work of mercy to alleviate human suffering,[74] which his critics thought a poor sort of humanity:

> If not a sparrow fillets to the ground,
> Without the notice of Almighty God,
> What will not be required of those who give
> Their sanction and support to such a crime
> As vivisection?
>   Not a throb or groan
> Of martyred animals strapped down in torture troughs
> (Within those 'cruel habitations' planned by cowards
> And human monsters known as lab'r'tries,
> Where 'science'—falsely called—holds unchecked sway,
> And cruelty un-masked stalks rampant in the midst
> Of dumb defenceless victims dazed with fear,
> And turning piteous eyes on the mean wretch
> Who stands, with Knife upraised, to make the gash
> Which is 'to benefit humanity'!)...[75]

If one can penetrate their execrable style, these verses neatly encapsulate the main argument against vivisection: it was inhumane and so could not benefit humanity, or be a path to knowledge, as only the morally ignorant would perform it.

The RDS seldom responded to religious or philosophical arguments, at times seeming genuinely baffled by them, and certainly did not rely on them to defeat the Dogs Protection Bill, arranging instead for their supporters in the House of Commons deliberately to prolong the proceedings so the bill would run out of time, a not uncommon parliamentary tactic.[76] The Bill, however, showed no signs of going away, and had yet another first reading in 1921.

At a public meeting of the London and Provincial Anti-Vivisection Society, one of its founders, the Irish suffragette Mrs Norah Dacre-Fox (later Norah Elam, 1878–1961), read out twenty letters from Members of Parliament in support of the Bill. As she knew that 'a large majority of the [female] public were strongly in favour of the measure', she felt sure it would pass 'if women made proper use of their new political power'.[77] There was, however, more power in influence than numbers, and the BMA's parliamentary subcommittee collaborated with the Commons medical committee to get the bill 'talked out' again, this time by the medical MP Francis Fremantle (1872–1943), who was acting 'on behalf of the [Research Defence] Society'.[78]

To forestall further attempts at legislation, Viscount Knutsford (the philanthropist Sydney Holland) requested that the BMA produce a definitive statement in favour of vivisection and they duly obliged, declaring in a memorandum of 1926 that any interference with it would 'impede advancement of knowledge'.[79] The following year, another Dogs Protection Bill, this time backed by the National Canine Defence League, was laid before parliament. By this time, public support had grown stronger, and the petition had a million signatures, including three thousand medical practitioners, coincidentally the same as the number of medical signatories on a petition in favour of vivisection that had led the Home Secretary to water down the provisions of the 1876 Act.[80]

The cardiologist Sir Thomas Lewis (1881–1945), who was said to have coined the term 'clinical science', hastily arranged for the BMA to convene a conference on research and animal experimentation, which predictably concluded that experiments on dogs were essential to 'the progress of medical science', and the bill was voted down at its second reading.[81] It did not help that the National Canine Defence League had overstated their case by claiming that physiologists were still staging 'demonstrations of a prolonged and agonising nature' for the benefit of their students, though such demonstrations were the one thing that public pressure had succeeded in curbing. The NCDL's out-dated caricature of medical teaching offended the physiologist J.B.S. Haldane (1892–1964), a humane man who became a vegetarian in later life and who maintained that scientists should avoid causing suffering to animals unless prepared to volunteer as experimental subjects themselves: he offered £100 reward for evidence of a cruel physiological demonstration having taken place within the last ten years, with no claimants.[82] The Dogs Bill was brought up again in 1925, only to be blocked in the Lords by peers representing the combined interests of the MRC, Royal College of Surgeons, Royal College of Physicians, and BMA.[83]

Throughout the inter-war period, the BMA staunchly opposed all parliamentary measures aimed at restricting vivisection.[84] In 1922, 1924 and 1930 Joseph Kenworthy MP (1886–1953) tried to introduce a bill on behalf of the BUAV to prevent National Insurance money (a form of income tax) being spent on vivisection, but the leaders of the BMA (without consulting their membership) rallied medical MPs to deny the bill parliamentary time.[85] Later in the year, the BMA's Secretary asked local branches to lobby their parliamentary candidates not to give the anti-vivisection pledge that some voters wanted. Included with the request was a list of MPs—67 out of a total of 615—whose anti-vivisection views were so well known that it was thought not worthwhile to approach them. All but six were members of the labour party.

One well-known socialist who did not agree with them was H.G. Wells (1866–1946), a graduate of the Royal College of Science in Kensington and sometime Labour parliamentary candidate for the University of London, who weighed in with a newspaper article denouncing anti-vivisectionists' 'fanatical illusions' and arguing that their real battle was not against cruelty but the scientific quest for knowledge.[86] Bernard Shaw replied for the antis that Wells's vision was the 'science' of imbeciles, since it would lead not to a better understanding of the world but to more and more introverted experimentation. The two writers personified the orthodox and alternative attitudes to science. To Wells's argument that the medical profession was 'massively in support of vivisection', Shaw replied that they had been taught to defend it as a 'tenet of faith', though they did not 'massively practice it'.[87] Though the RDS made much of the overwhelming support for vivisection amongst doctors, it was unsurprising given that no one could go through medical school without being indoctrinated in the importance of animal research.[88] Furthermore, for a doctor publically to support the anti-vivisection lobby was tantamount to professional suicide. The BUAV president Dr Walter Hadwen was barred from joining the BMA, and was subjected to what appears to have been a vexatious trial for medical manslaughter after the death of a patient in 1924.[89]

The repeated thwarting of the Dogs Protection Bill shows the strategic effectiveness of mobilising medical and parliamentary influence in support of animal experimentation. The RDS had less money to spend than the anti-vivisectionists, no donations from the public, and little popular support: it would have been impossible for them to muster a substantial petition *in favour* of experimenting on dogs, stray or otherwise.[90] However, they were able to persuade most of the few dozen medically and scientifically trained members of the House of Commons to oppose anti-vivisection bills whenever they arose. That these elected representatives had no qualms about ignoring public opinion reflects the paternalistic nature of medical science, as well as politics, at the time. Perhaps they decided that animal experimentation was for the good of the British people, whether they wanted it or not, though those MPs with connections in the research industry might be suspected of self-interest. Whatever their motives, pro-vivisection parliamentarians had the distinction of ignoring some of the largest petitions ever presented to the British government.

The involvement of the BMA was important in persuading both politicians and public that the nation's health and prosperity depended upon animal research. Led by a generation of doctors trained to accept laboratory experiments as the basis of medical knowledge, the

Association treated any threat to vivisection as a threat to their profession, which they countered by producing, on demand, pro-vivisection statements to suit the RDS's purposes.[91] In common with the RDS and the anti-vivisection lobby, the BMA did not limit its statements to answering questions of fact, but gave strong ideological support to a position that it regarded as non-negotiable.

'Dog Burke and Hare'

In a 1927 memorandum, the MRC stated 'There is no medical practitioner who does not use in his daily work information which he owes to experiments on dogs', and went on to say that, in many respects, the dog's anatomy was the nearest 'available' to that of man.[92] Whether this latter statement can be regarded as true depends on the significance of the word 'available'. It was certainly not the case that, as Viscount Knutsford told the House of Lords in 1924, the dog 'is more closely allied to man in what I may call its internal arrangements than is any other animal'.[93] Apes are obviously more closely related, and even if we charitably suppose his lordship meant British, domesticated animals, for a closer match he would have had to look no further than the pig. The truth was that experimenters preferred to use dogs because they were a convenient size to work with, relatively compliant, and so numerous they could be obtained cheaply and easily. The BMA went beyond defending the sale of dogs for experimentation, by demanding, in an echo of anatomists' calls for pauper dissection a century earlier, that the law be changed so that all strays that were 'unclaimed and obviously unwanted' were automatically made available.[94]

University College London was among the dog dealers' best customers. Its professors included some of Europe's most distinguished physiologists, whose students were exposed to a diet of experimental physiology far in excess of anything they needed to learn medicine: by the 1940s they were receiving a total of over 300 h of practical teaching in experimental physiology, around ten times more than in any present day medical school.[95] Obtaining sufficient animals for research and teaching on this scale was challenging, and while London, like any big city, had plenty of stray dogs, it was illegal under the 1906 Dogs Act to give or sell them to a laboratory. Dogs for vivisection had to be purchased from dealers, though as the provenance of any given dog was almost impossible to establish, the dealers found it easy to flout the law.

University College was first linked with dog stealing in 1913, when Professor Starling was summoned to the High Court to give evidence in a case.[96] Starling was a robust defender of the use of stray dogs for experimentation, arguing that as they were commonly euthanised anyway, they may as well be employed for useful purposes first.[97] On this occasion, the College was acquitted of any wrongdoing, and its physiologists continued to source dogs from local dealers. Thirteen years later, when sentencing a dog-stealer to six months hard labour for receiving and ill-treating two Irish terriers, a London magistrate alluded to the College's continuing involvement:

> You [Hewett the dog seller] are no doubt a cruel and unscrupulous man, and anything I can do to stop this sort of thing I will. I must not say too much because the people who employ you are not here and are not represented. Anyone who has heard this case must have a feeling of considerable uneasiness as to what is taking place. I have been told that a dog-stealer is employed by this school [University College] to supply them with dogs for physiological experiments.... It has been often said in these Courts that if there were no receivers there would be no thieves. At 8 a.m. two valuable pedigree dogs are missed from outside a house. At 9 o'clock they are taken to this school in a sack under circumstances of great cruelty, and in 24 h they would have been dead. No questions would have been asked. It must raise a feeling of considerable alarm among animal lovers to find that this has been going on for some time.

It certainly raised alarm, but it was impossible to prove that the physiologists knew the dogs they were buying were stolen. In their defence, the College pointed out '[t]hat the man Hewett has never been an employé ... [t]hat the professor of physiology had no means of ascertaining that Hewett had been convicted [in the past] of dog stealing', and that the professor had always '... required a written guarantee that all the animals so delivered by Hewett and by the other dealer with whom he traded were legitimately obtained'. The National Canine Defence League was suspicious: if the College was obtaining the dogs legally, why were they being brought there 'in sacks, as if they were potatoes'?[98] Lord Banbury thought the insistence on a written guarantee was also incriminating, since only a person who suspected they might be buying stolen goods would be sure to obtain one.[99] Moreover, the dogs showed signs of having been injured, as if their captors had found it necessary to subdue them.[100]

Rumours persisted, and University College soon found itself in the police news again, after one George Phipps was charged with stealing a wolfhound from outside its own home. The dog's 76-year-old owner tellingly went straight to UCL, where he inquired for Professor Ernest Verney (1894–1967). The Professor had the cages checked and the old man was reunited with his dog, which apart from a bump on the head was 'none the worse for his adventure'. At Phipps's trial (which the dog attended) there was more bad publicity for the College:

> A boy of fourteen, who said he was 'animal attendant' at University College, said he had known the defendant for about four weeks. He (the defendant) helped a man named Jackson to fetch dogs to the college.
>
> Counsel: Mr. Jackson often supplies dogs for the college?—Yes.
>
> He brought two on November 19?—Yes'.
>
> I would remark here that it is curious that a boy of fourteen should be employed to look after animals. I do not suppose any of your Lordships would give the charge of your animals solely into the hands of a boy of fourteen. But this is what emerges from those two statements, that within a fortnight two cases of stolen dogs are brought forward and in both of those cases these dogs were going to University College.[101]

The 'University of London Animal Welfare Society', set up by Starling to demonstrate that UCL took a responsible approach to research,[102] sometimes had to arrange for dogs to be nursed back to health to make them fit enough for vivisection, but had never questioned the vendors about why they were delivered in such a poor state.[103] Lord Banbury's allegations of a cover-up seem to have been warranted. When one medical MP commended the use of dogs in research on the grounds that they were cheap, this suggested, said Banbury, that they were being supplied illegally: 'Of course if you steal the dogs you do not pay much for them'.

In 1926, the BUAV decided there was enough evidence to fund an action against Verney, and while they were no doubt gratified when the sensational news that a University Professor had been summoned for 'receiving' was splashed across the newspapers, it soon became obvious that there was no chance of a conviction.[104] At Clerkenwell Police Court, Verney's innocence seemed to be assumed from the outset: he was allowed to sit at the solicitors' table rather than in the dock, and although the court was told the BUAV had paid for the plaintiff's lawyer, there was no mention that Verney's defence was supported by the RDS.[105] Dismissing the case, the magistrate said it should never have been brought, and ordered the prosecution to pay costs, though he did add that the College (which was buying over five hundred dogs a year) should make more stringent enquiries in future.[106] A spokesman for the College told the press: 'I am speaking for a large body of opinion which is tired of this slobbering by people who have nothing better to do than look after pups, parrots and pigeons'.[107] For the public, the message was simple: 'Watch Your Dog'.[108]

The parallels with body-snatching are extensive: the clandestine but widely-known market for 'subjects', the legal ambiguities, the professional denial of any suspicion, and the prosecution of middle men while the doctors went free. The defence of the physiologists who purchased dogs was the same as that of the anatomists who had purchased cadavers: they did not *know* that any crime had been committed to supply their needs, and were not responsible for the actions of others. Their shady deals were only possible with the complicity of a public most of whom simply did not care where scientists obtained their *materiel*. The animal victims, like the human victims of Burke and Hare (a comparison made in the press), were worth more dead than alive; they were, as one dog stealer told Starling's protégé Professor Lovatt Evans (1884–1968; his contribution to the war effort included working on poison gas at Porton Down), 'not worth a penny as dogs',[109] and like the victims murdered for dissection, they came mostly from the lowest classes, and nobody missed them. 'We want only mongrel dogs', said Evans,

'...valuable dogs would be too delicate for us': it required hybrid vigour to be vivisected.[110]

The BUAV would scarcely have been so naïve as to have expected a conviction; their motive was presumably to cause a scandal, and in this they succeeded, for even after Verney was acquitted on the legal technicality that he had not actually been in possession of the dog whilst it was in a cage in his department, the name of University College was still in the news for all the wrong reasons.[111] Questions were asked in the House of Commons and it made headlines that the College had 'used' 1,147 dogs in the past two years.[112] The inevitable public reaction followed, and the College received a flurry of letters: from anti-vivisection ladies, berating them for callousness and threatening divine retribution ('I am sorry for all vivisectors when their time comes to leave this world!'); from the owners of lost dogs, pleading for the professors to look in their laboratory cages; from people offering to sell unwanted dogs to the physiologists; and even one from a lady offering to sell her own body for research. One man who had sent a puppy to the vet to be destroyed only to discover that the lad who had taken it had sold it to a dealer 'for the sake of the money he gets from the Hospitals for vivisection' pleaded to be allowed to buy the dog back, to spare her further 'misery'.

Such compassion was lost on most experimenters: why, asked the RDS, did the anti-vivisectionists not simply accept the use of strays and so put an end to dog-stealing?[113] The fate of strays was, after all, a grim one. The RSPCA, the largest provider of homes for stray dogs, refused to sell them for vivisection, but could not cope with the numbers and destroyed tens of thousands every year by 'painless' electrocution.[114] Was not selling them to laboratories instead the logical thing to do? When Walter Hadwen, one of the few doctors still campaigning against anti-vivisection in the inter-war years, challenged the RDS about the vivisection of strays, they denied any knowledge of it.[115] They were in fact trying to get it legalised, and used their influence to plant a parliamentary question on the subject, having already supplied the Home Secretary with 'the necessary facts to provide an answer'. According to Lovatt Evans, assisting with a scheme to make stray dogs available for vivisection was 'the best opportunity that the RDS has ever had to render us [UCL physiologists] real service'.[116]

Conclusion

One of the most frequent criticisms of anti-vivisectionists between the wars was that they were motivated by sentiment and not logic; a criticism based, I have argued, on a paradigm of dispassionate, amoral science which, though prevalent, was still far from commanding universal assent. It might have been expedient to vivisect strays, as it had been to dissect paupers, but was it right? There were many in the anti-vivisection lobby whose feelings told them it was not. Vivisectionists, however, wielded influence where it counted. It was practically impossible to join the staff of a large teaching hospital if one was opposed animal experiments, and it was from this metropolitan élite that the leaders of the medical profession—presidents of the medical royal colleges and directors of research institutes—were drawn. They pronounced with authority that vivisection was necessary for medical progress, and it was difficult for laypeople or rank-and-file doctors to gainsay them.

Despite its influential supporters, the RDS saw itself as outnumbered and beleaguered by anti-vivisection campaigners with superior numbers and resources. The antis certainly had more money to spend, though any advantage was partially nullified by divisions within the movement, and winning the moral argument proved easier than winning the battle. It is significant that the RDS quickly abandoned their tactic of sending speakers to public meetings and engaging in competitive debates because discussion seemed to be getting them nowhere. It was easier to rely on bullying and intimidation: at a meeting in 1927, in the wake of the UCL dog-stealing scandal, Shaw was unable to make himself heard over the din of two hundred medical students, and in 1929, rowdy students literally broke up the annual general meeting of the BUAV.[117]

The fight for effectively unrestricted vivisection was won in the courts and parliament by clever tactics and collusion between those with vested interests. It is a moral certainty that the staff of University College knew that some of the dogs they purchased were stolen, but the RDS's lawyers correctly argued that as the physiologists had not technically been in possession of the dogs, they could not be guilty of receiving stolen goods. Parliamentarians knew that the public was opposed to vivisection, but the RDS and BMA could count on the support of enough members with medical interests to ensure that legislation to curb it was blocked at every stage, in the knowledge that, since animal experimentation was believed to contribute to national prosperity and security, no government would want anti-vivisection legislation to become law.

With legal challenges to vivisection blocked by parliament and the courts, and a mood of optimism that looked to scientific progress to bring peace and prosperity, anti-vivisection was beginning to look like a lost cause whose supporters were reactionary and selfish, putting their personal feelings before the interests of their own species and their own country. It would take the great depression of the 1930s to revive the link between radical politics and animal welfare, as the state's (mis)treatment of animals once more became a surrogate for its failure to protect its own citizens.

Notes

1. 'A member of the Provisional Committee' (letter), *BMJ*, *1* (1882), 599.
2. A. Flexner, *Medical Education in Europe: A Report to the Carnegie Foundation for the Advancement of Teaching* (New York: Carnegie Foundation, 1912), 120–7 (Flexner 1912).
3. A.W.H. Bates, 'Anatomy on trial: itinerant anatomy museums in mid-nineteenth century England', *Museum History Journal*, *9* (2016), 188–204 (Bates 2016).
4. Emma Hopley, *Campaigning Against Cruelty: the Hundred Year History of the British Union for the Abolition of Vivisection* (London: BUAV, 1998), 12–13 (Hopley 1998).
5. Tansey, 'The Queen'.
6. 'The Home Office report on experiments on living animals performed in 1906', *Lancet*, *1* (1907), 1502.
7. Kean, *Animal Rights*, 104.

8. Sometimes misreported as Harvey Cushing of the Johns Hopkins Hospital in Baltimore, who was a great friend of the RDS and wrote in 1913 that 'I may have said more than was necessary in regard to experimentation upon animals and the handicap which has been put upon British medicine by ill-advised legislation, but the subject is not one for silence if speech can in any measure lend encouragement and support to the work of the Research Defence League...' Harvey Cushing, *Realignments in Greater Medicine...* (London: Henry Frowde, 1914), 5 (Cushing 1914).

9. Minute book, 27 January 1908, Well SA/RDS/C1.

10. Memorandum from Paget, 21 May 1917, Well SA/RDS/C1.

11. C.S. Sherrington, 'Paget, Stephen (1855–1926)', rev. M. Jeanne Peterson, *Oxford Dictionary of National Biography* (Oxford: Oxford University Press, 2004). http://www.oxforddnb.com/view/article/35360, viewed 9 Aug 2015 (Sherrington 2004).

12. Hopley, *Campaigning Against Cruelty*, 10.

13. Kean, *Animal Rights*, 93.

14. The best history of hydrophobia in England dodges the question by refusing to countenance retrospective diagnosis, but the work of Pasteur and others cannot be properly evaluated without considering whether their diagnoses of hydrophobia were rabies or something else: Neil Pemberton and Michael Worboys, *Mad Dogs and Englishmen: Rabies in Britain, 1830–2000* (Basingstoke: Palgrave Macmillan, 2007) (Pemberton and Worboys 2007).

15. Austin Bradford Hill and Brian Hill, 'The life of Sir Leonard Erskine Hill FRS (1866–1952)', *Proceedings of the Royal Society of Medicine*, *61* (1968): 307–316 (Hill and Hill 1968).

16. 'Obituary: Charles Edward Beevor, MD, FRCP Lond', *BMJ*, *2* (1908), 1785–1786.

17. *The Fight Against Disease*, n.d. [1921], 11.

18. Hopley, *Campaigning Against Cruelty*, 23–5.

19. Minute book, 21 May 1908, Well SA/RDS/C1. In 'old' money there were 12 pence (d) in a shilling and 20 shillings (s) in a pound. Half a crown was 2s 6d.

20. Well SA/RDS/D1.

21. Chen-hui Li, 'Mobilizing Christianity in the anti-vivisection movement in Victorian Britain', *Journal of Animal Ethics*, *2* (2012), 141–61 (Li 2012).

22. Kensington branch Minute book, 17 May 1910, Well SA/RDS/C16.
23. Minute book, 23 June, 19 August, 20 October 1908, Well SA/RDS/C1; RDS, *The Fight Against Disease*, n.d. [1921], 12.
24. Minute book, 5 and 11 January 1909, Well SA/RDS/C1.
25. *Animals Defender and Zoophilist, 20/21* (1900), 165.
26. Richard D. Ryder, *Victims of Science: The Use of Animals in Research* (London: Davis-Poynter, 1975), 22 (Ryder 1975).
27. 'Anti-vivisection scene', *Daily Mail*, 6 October 1910 (Anti-Vivisection Scene 1910).
28. Minute book, 22 March, 17 May 1909, Well SA/RDS/C1.
29. Minute book, 19 July 1909, 24 October 1928, p. 119, 24 October 1934, p. 159, Well SA/RDS/C1; SA/RDS/C2; Kean, *Animal Rights*, 149.
30. Minute book, 1 February, 5 July 1909, Well SA/RDS/C1.
31. Minute book, 18 July 1910, Well SA/RDS/C1; Niven, *History of the Humane Movement*, 89.
32. Minute book, 19 July and 14 November, 1910.
33. Hopley, *Campaigning Against Cruelty*, 16.
34. Ulf Schmidt, 'Justifying chemical warfare: the origins and ethics of Britain's chemical warfare programme, 1915–1939', in David Welch and Jo Fox (eds) *Justifying War: Propaganda, Politics and The Modern Age* (Basingstoke: Palgrave Macmillan, 2012), 129–158, on pp. 140–141 (Schmidt 2012).
35. 'Experiments on living animals', *Lancet, 2* (1916), 440.
36. 'Experiments on living animals', *Lancet, 1* (1902), 1617–1618.
37. M. Beddow Bayly, 'The ill-effects of vaccines', *Animals' Defender, 60* (1941), 95 (Bayly 1941).
38. 'Anti-Vivisection in the war', *The Fight Against Disease* (1930), 10–11; Hopley, *Campaigning Against Cruelty*, 26.
39. 'Physiology during the war', *The Fight Against Disease*, January 1922, 2.
40. 'Anti-Vivisection: an enemy of the people', *The Fight Against Disease* (1927), 1–3.
41. RDS minutes, 19 February 1924, 59, Well SA/RDS/C2; *The Fight Against Disease* (1923), 1.
42. Letter from Paget to the RDS, 20 October 1920, Well SA/RDS/C2.
43. 'Notes and comments', *The Fight Against Disease* (1925), 11.
44. Debbie Tacium, 'A history of anti-vivisection from the 1800s to the present: part II (1914–1970)', *Veterinary Heritage, 31* (2008), 21–5 (Tacium 2008).

45. Schmidt, 'Justifying chemical warfare' (Schmidt 2012).
46. Hopley, *Campaigning Against Cruelty*, 27.
47. *The Fight Against Disease*, n.d. [1921], 2–3.
48. Carsten Timmermann, *Weimar medical culture: Doctors, Healers, and the crisis of medicine in interwar Germany, 1918–1933*. PhD, University of Manchester (1999) (Timmermann 1999).
49. Hopley, *Campaigning Against Cruelty*, 9.
50. Ryder, *Victims of Science*, 221.
51. RDS minutes, 24 October 1928, p. 119, Well SA/RDS/C2.
52. Robert Pearce and Warren Barr, *Pearce & Stevens' Trusts and Equitable Obligations*, 6th edn (Oxford: Oxford University Press, 2015), 404–405. It is interesting to speculate on whether this decision would be upheld today, now the importance of wilderness is better appreciated (Pearce and Barr 2015).
53. Memo from G.P. Crowd, honorary secretary RDS, 9 October 1934, Well SA/RDS/C2.
54. RDS minutes, 19 March 1931, p. 135, Well SA/RDS/C2.
55. RDS minutes, 11 March 1932, p. 145, Well SA/RDS/C2.
56. Memo from G.P. Crowd, 9 October 1934, Well SA/RDS/C2.
57. Vyvyan, *In Pity*, 9. Recent evolutionary theories suggest that the relationship is a biologically special one. For an overview of the coevolution of humans and dogs, see Raymond Coppinger and Lorna Coppinger. *Dogs: A Startling New Understanding of Canine Origin, Behaviour, and Evolution* (New York: Scribner, 2001) (Coppinger and Coppinger 2001).
58. 'Exportation of stolen dogs', *Bell's Life in London and Sporting Chronicle*, 28 May 1843, 1 (Exportation of Stolen Dogs 1843).
59. 'Dog's protection bill', *Nottingham Evening post*, 13 December, 1906, 6 (Dog's Protection Bill 1906).
60. House of Commons debate, 17 March 1905, *Hansard*, *143*, c395.
61. 'The vivisection controversy', *Sheffield Evening Telegraph*, 28 March 1906, 4 (The Vivisection Controversy 1906).
62. Hopley, *Campaigning Against Cruelty*, 16–17.
63. 'Problem for speaker', *Manchester Courier and Lancashire General Advertiser*, 17 July 1913, 5; 'Protection of dogs from vivisection', *Times*, 17 July 1913, 3 (Problem for Speaker 1913; Protection of Dogs from Vivisection 1913).
64. *Derby Daily Telegraph*, 24 July 1913, 2.

65. G.G. Greenwood, 'Sympathy and practice', *Times*, 24 July 1913, 14 (Greenwood 1913).

66. 'Experiments on dogs: medical protest against the prohibition bill', *Times*, 21 April 1914, 3 (Experiments on Dogs: Medical Protest Against the Prohibition Bill 1914).

67. 'Operations on dogs', *Aberdeen Journal*, 11 May 1914, 6.

68. 'London letter', *Western Daily Press*, 22 March 1919, 4; 'The vivisection of dogs: a reply', *The Fight Against Disease* (1924), 9; 'Medical research and the war', *Aberdeen Journal*, 28 June 1919, 5 (London Letter 1919).

69. 'The vivisection of dogs: a reply'; 'Dogs' protection bill', *Yorkshire Evening Post*, 27 June 1919, 8 (The Vivisection of Dogs: A Reply; Dogs' Protection Bill 1919).

70. *Western Daily Press*, 17 March 1920, 6.

71. 'The outbreak of rabies', *Times*, 23 April 1919, 13 (The Outbreak of Rabies 1919).

72. 'The dogs' protection bill', *BMJ*, *1* (1919), 613–4; Letter from Arthur H. Jona, *Hastings and St Leonard's Observer*, 12 May 1928, 10 (Jona 1928).

73. Well SA/RDS/D1.

74. 'Bishop and science: reply to anti-vivisectionists', Kensington branch minute book, 27 May 1910 [press cutting pasted in], SA/RDS/C16.

75. 'A new sort of postcard', *The Fight Against Disease* (1922), 9.

76. RDS minutes, 15 March 1920, 3, Well SA/RDS/C2.

77. Founded as the London Anti-Vivisection Society in 1876, it became the LPAVS in the early twentieth century, though there were never any branches outside London.

78. RDS minutes, 21 March 1923, 47; 22 May 1928, 113, Well SA/RDS/C2; *The Fight Against Disease*, n.d. [1921], 4.

79. BMA files, Well MP214 SA/BMA/C87.

80. Charles R. Johns, 'Petition to parliament', *Cornishman*, 9 March 1927, 7; Hopley, *Campaigning Against Cruelty*, 5; Theodore A. Cook, 'Dogs for research', *Times*, 7 March 1927, 20 (Johns 1927; Cook 1927).

81. Arthur Hollmann, *Sir Thomas Lewis: Pioneer Cardiologist and Clinical Scientist* (London: Springer-Verlag, 1997) (Hollmann 1997).

82. Ronald Clark, *J.B.S: The Life and Work of J.B.S. Haldane* (London: Bloomsbury Reader, 2011) (Clark 2011).

83. 'House of Lords, March 31st, 1925', *The Fight Against Disease* (1925), 13–14.

84. In 1927, the Secretary of the RDS wrote to the MRC to reassure them that, every year, the BMA had 'taken active steps in Parliament'

to prevent the passage of the various Dogs Protection bills, Well SA/BMA/C86.

85. 'Anti-vivisection bill', *Gloucester Journal*, 6 December 1930, 12; 'Vivisection bill opposed in public interest', *Western Daily Press*, 15 December 1930, 5 (Vivisection Bill Opposed in Public Interest 1930).

86. H.G. Wells, 'The way the world is going', *Sunday Express*, 24 July 1927 (Wells 1927).

87. George Bernard Shaw, 'These scoundrels!', *Sunday Express*, 7 August 1927 (George Bernard Shaw 1927).

88. 'Notes', *The Fight Against Disease* (1927), 30–1.

89. Hopley, *Campaigning Against Cruelty*, 37–42.

90. Letter from the medical secretary to the MRC, 11 March 1927, Well SA/BMA/C86.

91. In some cases, BMA members obviously did not support their Council's position: John Round was both honorary physician to the Anti-Vivisection Hospital and chairman of the BMA's Greenwich division: Obituary, *BMJ*, *1* (1937), 1234.

92. 'What mankind owes to dogs', *Daily Chronicle*, 8 June 1927, 4.

93. House of Lords debate, 25 March 1924, *Hansard*, *56*, c988.

94. Letter from the medical secretary to the MRC, 11 March 1927: Well SA/BMA/C86.

95. Tansey, 'The Queen'.

96. UCL archives, Kew (hereinafter, UCL) MSS add. 273.

97. Ernest Starling, 'On the use of dogs in scientific experiments', n.d., Well SA/RDS C1.

98. 'Echo of dog-stealing case', *Dundee Evening Telegraph*, 2 December 1926, 6; 'Vivisectors and dog-thieves', *The Abolitionist*, 1 December 1926, 139.

99. House of Lords debate, 8 December 1926, *Hansard*, *65*, cc1346–7.

100. 'Magistrate on sales in the Caledonian market', (newspaper cutting), UCL Bayliss papers, MSS add. 273.

101. Hansard, 8 December 1926, *65*, c1345.

102. 'Who sups with the Devil...', *Anti-Vivisection Journal*, June 1926, 71.

103. M.R. Ellinger, 'Antivivisection' (newspaper cutting), 24 December 1926, UCL MSS Add 273.

104. 'Vivisection: professor charged with receiving', *Johannesburg Star*, 9 December 1926.

105. 'Dogs for college experiments', *Dundee Evening Telegraph*, 22 December 1926, 5.

106. 'Summons against a professor dismissed', *Western News*, 23 December 1926, 2; 'Dogs for experiments', *The Star*, 14 December 1926 (Summons Against a Professor Dismissed 1926; Dogs for Experiments 1926).
107. 'Alarm among dog lovers', *Daily Herald*, 23 November 1926 (Alarm Among Dog Lovers 1926).
108. 'Watch your dog', *John Bull*, 4 December 1926.
109. 'Dog Burke and Hare', *Evening Standard*, 22 November 1926; Robert Blatchford, 'Betraying dumb friends', *Sunday Chronicle*, 28 November 1926, 10 (Burke and Hare 1926; Blatchford 1926).
110. 'The dogs a college bought from a thief', *Evening News*, 22 November 1926 (The Dogs a College Bought from a Thief 1926).
111. 'The dogs' case', *The Fight Against Disease* (1927), 3–11.
112. 'Dogs for experiments: 1,147 used at University College in two years', *Star*, 14 December 1926.
113. 'A challenge to anti-vivisectionists', *The Fight Against Disease* (1927), 15–16.
114. 'Stolen dogs case', *Daily Mirror*, 23 November 1926 (Stolen Dogs Case 1926).
115. RDS minutes, 7 December 1927, p. 107, Well SA/RDS/C2.
116. Memo from Lovatt Evans, November 1926, Well SA/RDS/C2.
117. 'Speakers howled down: uproar at anti-vivisection meeting', *Daily Mail*, 17 June 1927; Hopley, *Campaigning Against Cruelty*, 43.

References

Anti-Vivisection Scene. (1910, October 6). *Daily Mail*.

Alarm Among Dog Lovers. (1926, November 23). *Daily Herald*.

Bates, A. W. H. (2016). Anatomy on trial: Itinerant anatomy museums in mid-nineteenth century England. *Museum History Journal, 9,* 188–204.

Bayly, M. B. (1941). The ill-effects of vaccines. *Animals' Defender, 60,* 95.

Blatchford, R. (1926, November 28). Betraying dumb friends, *Sunday Chronicle*, 10.

Burke, D., & Hare. (1926, November 22). *Evening Standard*.

Clark, R. (2011). *J.B.S: The life and work of J.B.S. Haldane*. London: Bloomsbury Reader.

Cook, T. A. (1927, March 7). Dogs for research. *Times,* 20.

Coppinger, R., & Coppinger, L. (2001). *Dogs: A startling new understanding of canine origin, behaviour, and evolution*. New York: Scribner.

Cushing, H. (1914.). *Realignments in greater medicine...* (p. 5). London: Henry Frowde.

Dogs for Experiments. (1926, December 14). *The Star.*

Dog's Protection Bill. (1906, December 13). *Nottingham Evening post,* 6.

Flexner, A. (1912). *Medical education in Europe: A report to the carnegie foundation for the advancement of teaching* (pp. 120–127). New York: Carnegie Foundation.

Experiments on Dogs: Medical Protest Against the Prohibition Bill. (1914, April 21). *Times,* 3

Exportation of Stolen Dogs. (1843, May 28). *Bell's Life in London and Sporting Chronicle,* 1.

George Bernard Shaw. (1927, August 7). These scoundrels! *Sunday Express.*

Greenwood, G. G. (1913, July 24). Sympathy and Practice. *Times,* 14.

Hill, A. B., & Hill, B. (1968). The life of Sir Leonard Erskine Hill FRS (1866–1952). *Proceedings of the Royal Society of Medicine, 61,* 307–316.

Hollmann, A. (1997). *Sir Thomas Lewis: Pioneer cardiologist and clinical scientist.* London: Springer.

Hopley, E. (1998). *Campaigning against cruelty: The hundred year history of the British Union for the abolition of vivisection.* London: BUAV.

Johns, C. R. (1927, March 9). Petition to parliament. *Cornishman,* 7.

Jona, A. H. (1928, May 12). *Hastings and St Leonard's observer,* 10.

Li, C. (2012). Mobilizing Christianity in the antivivisection movement in victorian Britain. *Journal of Animal Ethics, 2,* 141–161.

London Letter. (1919, March 22). *Western Daily Press,* 4

Pearce, R., & Barr, W. (2015). *Pearce & Stevens' Trusts and Equitable Obligations* (6th ed., pp. 404–405). Oxford: Oxford University Press.

Pemberton, N., & Worboys, M. (2007). *Mad dogs and Englishmen: Rabies in Britain, 1830–2000.* Basingstoke: Palgrave Macmillan.

Problem for Speaker. (1913, July 17). *Manchester Courier and Lancashire General Advertiser,* 5.

Protection of Dogs from Vivisection. (1913, July 17). *Times,* 3.

Ryder, R. D. (1975). *Victims of science: The use of animals in research.* London: Davis-Poynter.

Schmidt, U. (2012). Justifying chemical warfare: The origins and ethics of Britain's chemical warfare programme, 1915–1939. In D. Welch & J. Fox (Eds.), *Justifying war: Propaganda, politics and the modern age* (pp. 129–158, on pp. 140–141). Basingstoke: Palgrave Macmillan.

Sherrington, C. S. (2004). Paget, Stephen (1855–1926), rev. Jeanne Peterson M., *Oxford dictionary of National Biography.* Oxford: Oxford University

Press. Retrieved August 9, 2015, from http://www.oxforddnb.com/view/article/35360.

Stolen Dogs Case. (1926, November 23). *Daily Mirror*.

Summons Against a Professor Dismissed. (1926, December 23). *Western News*, 2.

Tacium, D. (2008). A history of antivivisection from the 1800s to the present: part II (1914–1970). *Veterinary Heritage, 31*, 21–25.

The Dogs a College Bought from a Thief. (1926, November 22). *Evening News*.

The Outbreak of Rabies. (1919, April 23). *Times*, 13.

The Vivisection Controversy. (1906, March 28). *Sheffield Evening Telegraph*, 4.

The Vivisection of Dogs: A Reply; Dogs' Protection Bill. (1919, June 27). *Yorkshire Evening Post*, 8.

Timmermann, C. (1999). *Weimar medical culture: Doctors, healers, and the crisis of medicine in interwar Germany, 1918–1933*, PhD. University of Manchester.

Vivisection Bill Opposed in Public Interest. (1930, December 15). *Western Daily Press*, 5.

Wells, H.G. (1927, July 24). The way the world is going. *Sunday Express*.

7

State Control, Bureaucracy, and the National Interest from the Second World War to the 1960s

In 1935, Norah Elam, the former Mrs Dacre-Fox—separated from her husband and now using the surname of her lover, Edward Elam—having recently joined, and risen to prominence in, the British Union of Fascists, published *The Medical Research Council, What It Is and How It Works*, a distillation of the information she had gleaned while working in the MRC typing pool during the Great War. In this pamphlet, she questioned whether animal research could safely be extrapolated to humans, and why so many experiments were either repeats, or else yielded results apparently obvious to anyone with common sense. She was not alone in thinking that the use of animal models for human disease had gone too far: some among the medical profession were complaining that laboratory experimentation had become the master rather than the servant of medicine, to the detriment of clinical studies.[1] Elam laid the blame at the door of 'powerful vested interests' (which, in the context of her political views, meant Jewry) that had managed to 'entrench' themselves behind 'State-aided research', where they could exercise control without being accountable to the public.[2]

© The Author(s) 2017
A.W.H. Bates, *Anti-Vivisection and the Profession of Medicine in Britain*,
The Palgrave Macmillan Animal Ethics Series, DOI 10.1057/978-1-137-55697-4_7

British Fascists and Anti-Vivisection

In the political climate of the depression years, right wing activists found common cause with anti-vivisectionists, exploited animals being to fascists, as they had once been to socialists, a symbol for the fate of downtrodden workers in a society where profit came before people, and where a shadowy oligarchy manipulated the poor for its own commercial ends. In Germany, to which other European proto-fascists looked to see their principles put into practice, the Nazis had banned vivisection in 1933, soon after coming to power, a popular move in a country enthusiastic for *Lebensreform*, and also with British anti-vivisectionists, who were of course unaware that the Nazi government would not scruple to sanction experimentation, on animals or humans, when it suited their purposes.

Putting an end to vivisection was for British fascists, as it had been for socialists and the new age movement, part of a utopian plan for liberating the oppressed and re-establishing the natural order, the latter being, as for all ideologues, the state of existence most congruent with their own politico-religious views. Elam was not the only animal welfare campaigner to embrace fascism—Maidie Dudley Ward (d. 1945) was active in the RSPCA, the Animal Defence and Anti-Vivisection Society, the Nordic League, and Oswald Mosley's January Club.[3] For anti-vivisection's critics, the link with fascism showed that the movement's supporters were fundamentally misanthropic: in trying to bring animals closer to the level of humans they were in fact reducing the lowest humans to the level of animals.[4]

British fascists were, however, a vocal minority who never came close to gaining power, though their involvement did bring some new life to organised anti-vivisection. Elam invited the former director of propaganda for Mosley's Blackshirts, Wilfred Risdon (1896–1967), to join her at the LPAVS, thereby introducing to the movement one of its most capable leaders, as well as drawing down upon it increased government scrutiny.[5] Soon after the outbreak of the Second World War, both Elam and Risdon were arrested by Special Branch, an action the police claimed was justified after a search of the LPAVS offices uncovered 'a list containing the names of eight members of the B[ritish] U[nion of Fascists] and a letter from Oswald Moseley [*sic*]'.[6]

It was partly on the basis of this evidence that the historian Richard Thurlow described the LPAVS as 'a known fascist front organisation'.[7] This was certainly how it was perceived by the authorities, and Elam made no secret of her political views in Society meetings (at least one other committee member was an active fascist), but to call the LPAVS a 'front' is an overstatement. The Society had been active in defence of animals since the beginning of the twentieth century, and Elam had been involved from the earliest days, for most of the time while a member of the Conservative party. As her politics became more extreme, she began to proselytise for the Blackshirts as a personal initiative, and found some fellow anti-vivisectionists sympathetic to the cause, but the LPAVS's anti-cruelty mission was genuine enough.[8]

As 'Nazi sympathisers', Elam and Risdon were imprisoned without trial under Defence Regulation 18B, but Risdon swiftly disowned the British Union of Fascists and was promptly released to return to his work at the LPAVS, of which he became secretary in 1942. In this capacity, he exchanged ideas with Air Chief Marshall Sir Hugh Dowding, the former head of RAF Fighter Command, and certainly no sympathiser with the enemy, whose innovative mind, freed from wartime responsibilities by premature retirement, found an outlet in various new age causes. Elam, however, remained in detention, casting a shadow over the LPAVS and rekindling the suspicion fomented during the First World War that anti-vivisectionists were incipient traitors. The LPAVS did their best to distance themselves from embarrassing political links by putting a notice on the front page of their news sheet assuring readers that none of their committee was a member of any 'suspect organisation'.[9]

An even greater problem for them was that, now war had broken out, vivisection seemed a comparatively trivial issue. In an editorial, they defended their continued activity on the grounds that firm moral principles were more important than ever in wartime, and that denunciations of Nazi cruelty would be hypocritical on the lips of those who were cruel themselves, though as Germany had stronger legislation to protect animals that any other country in Europe, the argument that anti-vivisectionists were never cruel seemed rather flimsy.[10] The LPAVS chairman, Captain Guy Coleridge, RN (1884–1941), patriotically tried to show that National Socialism was not the panacea for animals that it seemed, but the best he could come up with was a bizarre story that Hitler had personally given orders that all dogs in Germany should be killed.[11]

The LPAVS's efforts to show they were not assisting the enemy were unfortunately nullified by their campaign against compulsory vaccination, to which their own and the BUAV's publications devoted an increasing amount of space. There had always been some opposition to vaccination from anti-vivisectionists on animal welfare grounds, for example the Anti-Vivisection Hospital's prohibition of vaccines prepared from live animals, but anti-vivisectionists were now opposing vaccination for libertarian reasons. Hadwen had always been an anti-vaccinationist—Cobbe had recruited him to the BUAV after hearing him speak at an anti-vaccination rally—as he felt that patients ought not to be forced into accepting scientific 'progress'.[12] The outbreak of war had given the government an excuse to impose vaccination on servicemen, but anti-vivisection groups failed to appreciate that, in objecting to what they saw as an experiment on unwilling soldiers they appeared to be interfering with the war effort and showing disloyalty to the national government, an unfortunate impression for a movement linked with pacifism and fascism to give.

The War Years

The impact of the War on laboratory animals was largely negative. There was an initial reduction in animal use as peacetime research projects were shelved for lack of funding, and in July 1944, German bombs achieved what anti-vivisectionists had long failed to do, closing down the infamous Brown Institute, but many additional animals were being sacrificed in military tests, and with private members' bills banned there was no chance of anti-vivisection MPs preventing this.[13] The last peacetime Home Office returns, in 1939, reported a total of 908,846 experiments in the previous year, but during the war only simplified reports were issued and the secrecy surrounding experimentation was increased, making it difficult to discover how many animals were used and what experiments were being performed on them. In the absence of reliable information, rumours abounded: the Ministry of Agriculture was said to be seeking 'unlimited numbers' of hedgehogs to 'help to win the war', while 'pet stores' advertised for ten thousand guinea pigs, whose contribution to the war effort remains a mystery.[14]

As the war progressed, reports of research carried out in government facilities began to appear in the medical press. The LPAVS reacted critically to a paper by Solly (later Lord) Zuckerman (1904–1993) in the *Lancet* describing a study of the effects of blast injuries on unanaesthetised mice, rats, guinea pigs, rabbits, cats, monkeys and pigeons, which had been placed as little as thirteen feet from seventy pounds of high explosive. Predictably, those that were not blown to bits mostly died from traumatic haemorrhage of the lungs.[15] In retrospect it is difficult to see what purpose these experiments served, since the animals chosen were generally too small for their injuries to be comparable to those of humans, and in wartime there were plenty of human fatalities in which the effects of blast injuries could have been studied at autopsy.

Many other experimenters sought to reproduce in the laboratory the traumas experienced by humans in war: researchers in the anatomy department at Oxford crushed guinea pigs' legs with metal rods and introduced bacteria into the wounds to make them suppurate; at University College Hospital, they burned goats and killed the survivors at intervals to study the pathology of their skin; and in the physiology department of King's College Hospital they administered fifty blows to the thigh bones of cats with a mallet, fracturing them in every case.[16] In all these experiments the animals were anaesthetised when their injuries were inflicted, but they were allowed to regain consciousness later, and some were subsequently experimented upon again. Though these researches were openly reported in medical journals in 1943 and 1944, it was 10 years before the LPAVS ventured to criticise them in print.[17] Other casualties of war included animals of various species that were exposed to poison gas, in anticipation of a gas attack on the British mainland that never occurred, and eighty sheep, infected during the secret testing of an anthrax bomb on the remote Scottish island of Gruinard.[18]

Many of the LPAVS's wartime initiatives were uncontroversial: Risdon designed an ingenious air raid shelter for domestic pets, and they continued to campaign against cruelty in the meat industry.[19] It was, predictably, their denunciation of the government's policies on vaccination and military experimentation that led to confrontation: the RDS reported the LPAVS to the Parliamentary Medical Committee as a 'malevolent influence', and asked the Committee chairman, Sir Francis Fremantle, to put 'pressure' on them.[20] The Army Director of Pathology concurred: having to answer questions about experiments and deal with complaints about compulsory vaccination (the government grudgingly had to admit that soldiers were free to refuse the vaccines if they chose) was a waste of army time and delayed more important work. 'In addition to jeopardising the safety of the individual soldier', he wrote, 'the activities of these [anti-vivisection] societies are a menace to the national effort at this time, I am, therefore of opinion that the strongest possible action should be taken at once to restrain their further activity'.[21]

Despite these forewarnings, six anti-vivisection societies rashly came together in 1942 to oppose compulsory diphtheria inoculation of troops.[22] This ill-timed move gave the treasury the excuse it needed to revoke their charitable status, a change to which Fremantle and the RDS lent their support.[23] The public had grown tired of anti-vivisectionists stirring up dissent in the ranks, and the RDS was pleased to note that 'our society has never done anything more popular'. They swiftly arranged for anti-vivisection societies to be removed from all published lists of charities.[24] When the war was over, there was no public appetite for reversing the decision: the House of Lords rejected a final appeal by the NAVS in 1947.[25]

The judgement was a severe blow for anti-vivisection, whose charitable status had been accepted since the foundation of the VSS in 1875, and upheld in court in 1895. Though it was unusual for charitable status to be granted to any organization whose objects included changing the law, the court ruled that the VSS's overarching purpose was to end what it saw as 'a cruel and immoral practice',[26] a goal it believed would benefit humans as well as animals (whether it would actually do so was not for the court to decide, it sufficed that the charity believed it would). In 1947, however, when the NAVS, as it then was, tried to get its charitable status restored, the Tax Commissioners argued that any benefit to public morals from the abolition of vivisection would be negligible in comparison to the damage to public health. According to the appeal court judge Lord Wright, 'the calamitous detriment of appalling magnitude' that would be suffered by medical science if vivisection were stopped greatly outweighed any 'vague and problematical moral elevation' that society might gain.[27] In their judgement, the Law Lords also adverted to the political nature of the Society's objective. It was a judgement based on materialistic utilitarianism, and it is difficult to see how anyone of this turn of mind could have dissented from it, but it totally disregarded the century-long debate about the nature and purpose of science, which had by, this time become so passé as to elicit little interest outside the dwindling ranks of committed anti-vivisectionists.

Post-War Problems

In the immediate post-war period, civil experimental programmes were resumed with such enthusiasm that the price of laboratory animals rose sharply owing to shortage of supply. In Bristol, physiologists were prepared to pay up to 17 shillings, a labourer's daily wage, for a cat, which suggests they were no longer being offered sufficient numbers of unwanted or stolen domestic animals and strays, probably because many of them had been euthanized during the war, purportedly in the national interest: 400,000 cats and dogs had been massacred in London alone in 1939 as the result of unfounded fears of wartime food shortages to come.[28]

As the already illicit supply of 'strays' to laboratories was insufficient to meet their demands, major consumers such as the pharmaceutical company Burroughs Wellcome found themselves purchasing, perhaps inadvertently, stolen pets. In a notable case in 1945, one Mr Bailey located his 'lost' dog, Digger, when he heard his distinctive barking coming from a crateful of dogs bound for the Wellcome laboratories in Bradford. An attempt by the BUAV to use this incident to publicise the illegal dog trade was blocked by threats of a libel action, and newspaper reports elicited remarkably little public concern: the argument that medical progress was impossible without vivisection was now generally accepted, and anti-vivisection groups had lost their political and charitable credibility.[29]

Details of Nazi medical experiments, when they emerged, only made things worse. Although Vyvyan has argued that all experiments on humans in Nazi Germany 'were in continuation of, or complementary to, experiments on animals',[30] the obvious interpretation of the fact that the most outspokenly anti-vivisection government in history had the worst record on human rights was that ostentatious concern for animals masked an underlying misanthropy in which the value of human lives was debased.[31] The priority in post-war Europe was to strengthen human rights, and the 1948 Universal Declaration of Human Rights concentrated the attention of moral reformers on issues such as judicial corporal punishment and the death penalty. Animals were not mentioned in the Declaration at all, and a comparable declaration of rights for them is still awaited.[32] In some respects, human rights and animal rights had become competing interests: it was apparent that improvements in living standards and health would be critical in preventing future conflicts, and state-sponsored medicine and animal research were seen as vital for achieving this.

In Britain, the National Health Service Act of 1946 was initially welcomed by anti-vivisectionists because they thought it would make it easier for patients to opt out of vaccination, although any who chose to do so were probably more concerned about potential side effects than the use of live animals. In fact, by reducing patient choice, state-run medicine tended to restrict patients' ability to exert moral influence. Although the right of patients to choose their doctor and doctors to choose their patients was enshrined in the Act, it was meaningless in practice because both groups had their freedoms curtailed under the nationalized system.

In 1948, anti-vivisection organizations, concerned that experimentation had become routine, regulation a formality, and dispensations from anaesthesia the norm, requested that the government set up another Royal Commission to revise the 1876

Act, but their plea was ignored.[33] The BMA staunchly defended experimentation and opposed any changes, insisting that all necessary safeguards were in place.[34] Clearly, however, the Act's effectiveness as a regulatory agent was highly questionable: when the BUAV asked in 1954 if the government had ever turned down a licence application, the Secretary of State replied that the information was not available, which suggests that they had not.[35] It was rumoured that the government's secret animal research programme was still going on, but questions in the House of Commons about whether animals were being used in American-style atomic weapons testing met with a wall of silence, as it was deemed 'not in the public interest' to answer them.[36] The director of Britain's atomic research establishment at Harwell did, however, admit that animals had been exposed to radiation, and apparently told a BUAV supporter that 'the end justifies the means'.[37]

LD50

The major change for laboratory animals in the 1950s was the same as that for the medical profession and the population as a whole: they became increasingly subject to state control. The great majority of research on animals was now a matter of bureaucratic necessity rather than, as it had been when the anti-vivisection movement began, the personal initiative of a few ground-breaking physiologists. Until the 1920s, research had been mostly qualitative, directed at determining how animals functioned and how they reacted to disease, either for academic interest or, more often, to provide a model for human pathophysiology. The total number of animals used in qualitative studies was comparatively small: the antis criticised unnecessary repeat experiments and demonstrations done purely for teaching purposes, but their main complaint was that vivisection was demoralising to those who performed it, to the profession of medicine, and to society as a whole.

Paradoxically, opposition to vivisection in the post-war period, when quantitative testing predominated, became less vocal although the number of animals used increased. This was due in part to experimenters having won the propaganda battle by convincing the public that they were saving lives and helping their country to prosper, while anti-vivisectionists were sentimental, reactionary, and disloyal to their own species. Also, Joseph Stalin's apocryphal dictum probably applied: 'a single death is a tragedy; a million deaths is a statistic'. A solitary physiologist choosing to vivisect a stolen dog in a private laboratory was more likely to provoke an emotive response than any number of routine tests carried out by white-coated technicians on anonymous animals that would never see life outside a laboratory.

Chief among these bureaucratised consumers of animals was the lethal dose test, which had originated in 1921, when Dr A.J. Eagleton (1891–1925) of the Wellcome Physiological Research Laboratories proposed a method to standardise the potency of tuberculin by measuring its 'minimum lethal dose' in guinea pigs.[38] It was the potency rather than toxicity of the vaccine that was in question, but the latter was a convenient proxy for the former, since it was harder to test the strength of a vaccine than to find the lowest dose that would cause death. The test was seriously flawed, because susceptibility to toxins varies both between and within species, and it took only one idiosyncratic result to skew the findings. This problem seemed, however, to have been resolved in 1927, when the Dr J.W. Trevan (1887–1956), who had taken up laboratory work because he 'found clinical medicine too difficult',[39] proposed measuring the dose necessary to kill half the population to which it was given, to so-called *dosis letalis* 50% or DL50, which was soon anglicised to LD50.[40] Trevan intended his method to be used for standardizing drugs such as digoxin and insulin that varied from batch to batch and were dangerous in overdose. He appreciated that this would require 'much larger' numbers of animals than minimum lethal dose testing, and made some suggestions for 'economy', though probably with financial rather than humane considerations in mind.

The potential for using animals to test the safety of medicines caught the attention of the BMA, who raised it during a debate on Joseph Kenworthy's bill to stop public money being spent on vivisection, arguing that animal testing was necessary to guarantee that medicines were safe, and that 'effective control of therapeutic substances can only be ensured by the state…'.[41] Though officially apolitical, the BMA, as we saw in the previous chapter, wielded significant parliamentary influence, not by 'retaining' (i.e., paying) MPs, but by persuading the medical men among them to ask planted questions, or block legislation by 'talking out' or delaying bills, in the knowledge that no government would give anti-vivisection extra parliamentary time.[42]

LD50 testing was little used—or at least little reported—in Britain until the Second World War. The first research published in the *Lancet* that employed the technique was a 1943 study, jointly funded by the MRC and Boots Pure Drug Company, into the toxicity of an unknown substance that had been extracted from dead muscle (the object being to investigate the systemic effects of soft tissue injuries).[43] The mystery compound was variously fed to, or injected into, the veins or abdominal cavities of unspecified numbers of cats, rabbits, rats, mice and guinea pigs. Not surprisingly, there were 'wide differences' in response, both within and between species. The 'extreme variability' of the rabbit, the investigators concluded, rendered it 'quite unsuitable for biological assays', a recommendation that the pharmaceutical industry would, in decades to come, comprehensively ignore.

A few weeks later, the professor of morbid anatomy at UCL published a report into the LD50 of tannic acid, a substance of interest to the War Office as it was being tried out in the treatment of burns. The experimenters used of a total of 250 goats, rabbits, guinea pigs and rats, a tenth of whose skin surface was burned off under anaesthesia before they were sprayed with the acid.[44] Wartime necessity allowed such experimentation to escape public censure, as it was intended to alleviate the sufferings of wounded combatants. Another paper in the *Lancet* in 1945 reported the efforts of the biochemistry department at Oxford to develop an antidote to arsenical gases, which, it was feared, might be deployed in a last-ditch German attack. Their LD50 was determined by applying them to the skin of rats, presumably causing considerable pain, since a human 'volunteer' who had as little as one milligramme rubbed onto his arm experienced oozing and redness.[45]

One reason that lethal dose testing did not generate a significant public reaction despite the large quantities of animals used and the suffering it caused was that rodents soon became the animals of choice. Although experimenters typically used multiple species to counter the problem that interspecies variation was wide, rodents were the default option. The era of the 'lab rat' may be said to have begun in 1909, when a standard strain, the Wistar rat, was bred specifically for experimental use—the ancestor of the majority of laboratory rats used thereafter. The advantages were readily apparent: rats have a conveniently short generation time, reproduce easily, and are seen by the public as vermin and thus engender little sympathy, particularly those bred for the laboratory that have never been wild animals or pets. It is notable that illustrated propaganda from anti-vivisection groups rarely depicted rats, whereas pro-vivisection literature often did. The massive breeding programmes necessary to provide them in the large numbers required also had the desirable side effect of reducing intra-species variability, since the population became unnaturally genetically uniform due to inbreeding.

By the 1950s, LD50 testing was responsible for most animal deaths in the laboratory, and for a huge rise in the total number of experiments carried out. Though toxicity tests did appear in official statistics, the government dissembled by calling them 'simple injections', without adding '…of fatal poisons'. In fact, LD50 was almost bound to produce the maximum suffering possible, since the target dose was one that was only just fatal, perhaps after many days. Unlike the vivisection experiments of the nineteenth century, which were often public and involved mostly domestic animals whose sufferings were easily anthropomorphised, laboratory animals were experimented on in private, with bureaucratic efficiency, and the results reported in such a way that the animals were hardly even mentioned. Phrases in the academic literature such as 'the LD50 was determined' glossed over hundreds of slow, painful deaths. This routine, industrialised killing of creatures, without regard for suffering, was carried out not by medical visionaries but anonymous technicians, to whom the attribution of motives either of brutality or nobility of purpose would have seemed equally redundant.

Such was the confidence placed in LD50 testing that it was extended beyond pharmaceuticals to a bewildering range of domestic products, chemicals and cosmetics, though in many cases testing these for toxicity seems to have served little purpose. The BUAV took up the test case of the insecticide DDT, which was tested on a variety of domestic animals. Since an insecticide must obviously be poisonous, and since DDT would not in the real world be given to either animals or humans, why, they asked, were the tests needed at all? D.W. Jolly, the veterinary surgeon in charge, replied glibly that it was necessary to test any potentially dangerous chemical, though he added that he personally disliked the work and was reluctant to perform it. In its defence, he produced not the classic utilitarian argument but the bureaucrat's customary excuse for any misdeed: the tests had, he said, been 'planned by a committee', thus, presumably, absolving him, and anyone else, from personal responsibility. The fact that the authorities 'demand' such tests, replied the BUAV, only showed how foolish the system was, since the results were easily predictable.[46] Their objection was, however, brushed aside, and examples of similar senseless experiments—from injection of known poisons on the one hand to determining the LD50 of water on the other—might be supplied in abundance.

Testing on laboratory animals had now won such widespread scientific endorsement that its value had become practically unquestionable, not least because many of the leading figures in academic biomedicine had built their reputations on repetitive, protocol driven, quantitative experimentation, and continued to support it. Trevan, the inventor of the LD50 test, became a Fellow of the Royal Society, Research Director at the Wellcome Laboratories, advisor to the government, and Chairman of the RDS. His former assistant George Alexander Mogey (1917–2003) was Secretary of the Council for Postgraduate Medical Education, where he commemorated his earlier career in the laboratory by acquiring the car registration plate 'LD50'.[47]

The Sacred Cow of Science

In 1953, the LPAVS's position on medical science was succinctly set out in a review of Anthony Standen's book *Science is a Sacred Cow*: '… Standen shows the sacred cow as an unimpressive figure when she has a halo round her horns and is surrounded by white-coated figures bowing low, but he also shows that she remains just as good a cow, and gives as nourishing milk, when we treat her properly in her barn or in her meadow'.[48] Most anti-vivisectionists did not disdain science, but argued against excessive reliance upon it, and in particular against the requirement for every discovery to be 'validated' by experiments on animals. They tended, however, to ignore the rodents that made up the majority of the victims and to concentrate on saving domestic animals, especially dogs, though their efforts to get a dogs protection bill through parliament remained ineffectual. The first case of the 'liberation' of laboratory animals occurred in 1952, when an anti-vivisectionist released eight dogs from the kennels of a dealer. Ironically, considering the number of 'strays' and 'lost' dogs that were being kidnapped daily to supply laboratories, the dog-rescuer was convicted of stealing them, though he was conditionally discharged. The incident prompted the BUAV to start a campaign to raise money to buy up unwanted dogs, and so prevent them falling into the hands of laboratory suppliers.[49]

In an attempt to heal some of the divisions that beset the movement, a 'World Congress' of anti-vivisection societies met in London in 1954, with Risdon in the chair. He was the closest thing that British anti-vivisectionists had to a national leader, and his propaganda experience proved valuable in maintaining their public profile. He tried to improve long term support in the House of Commons by asking LPAVS members to 'badger' prospective parliamentary candidates about animal welfare issues, and contributed to radio discussions whenever he could, though he felt the BBC was biased in favour of vivisection and uncritically presented the government's position as authoritative.[50]

Chief among a dwindling number of anti-vivisection parliamentarians was Hugh Dowding, who had become a theosophist since his elevation to the House of Lords in 1943. In 1952, in a speech against animal experimentation, he rejected out of hand the defence of utility: '… even should it be conclusively proved that human beings benefit directly from the suffering of animals, its infliction would nevertheless be unethical and wrong'. In 1957, he attacked the secrecy surrounding animal research and summed up the regulatory system with military bluntness: '… a hollow sham, maintained to throw dust in the eyes of critics and to salve the conscience of the apathetic'.[51] His renewed calls for a government enquiry fell on deaf ears in parliament, and the RSPCA and Universities Federation for Animal Welfare

(UFAW, a graduate-only anti-vivisection society) added their voices to the appeal with scarcely more success. The RSPCA tried to meet the Home Secretary to tell him that five inspectors for millions of experiments was clearly inadequate, but he refused to receive their delegation. They also produced a leaflet, *Cruelty Within the Law*, which pointed out that the licensed experiments performed without anaesthetic included starvation, inoculation with virulent diseases, sleep deprivation, and exposure to poison gas. The Home Secretary's only concession was to appoint a sixth inspector—another doctor rather than the veterinarian the RSPCA had requested.[52]

Major C.W. Hume, founder and chairman of the UFAW, delivered a keynote speech in (1958) that stressed the historic virtue ethics argument, comparing experimenters who were thoughtless in their use of animals to First World War generals who coldly sacrificed their troops. Even if the latter's actions did ultimately lead to military victory, which must be the prime objective of any commander, their callous indifference to life would still have been wrong, on the grounds of both inhumanity and inefficiency. Hume, who was perhaps mindful, as a soldier, that the most reckless of commanders were often those who faced no personal risk, criticised experimenters for sacrificing animals for what they insisted were worthy causes, and yet declining to make any experiments upon themselves. The *Lancet* reprinted the speech with approval, exhorting experimenters to be more efficient and to reduce suffering whenever they could.[53]

Wishing 'to see laboratory techniques become more humane for the animals concerned', the UFAW commissioned the Oxford zoologist Dr William Russell to undertake a thorough study of the subject. Russell and his assistant Rex Burch published the results of their several years' work in (1959), as *The Principles of Humane Experimental Technique*, an influential report most notable for proposing the so-called 'three Rs': Replacement, Reduction, and Refinement of animals in laboratory experimentation.[54]

The NAVS, meanwhile, preferred to fight utilitarian science on its own terms by arguing that vivisection was not necessary for effective medical research. In the 1960s it published a series of short books to this effect by the theosophist and anti-vivisection doctor Maurice

Beddow Bayly (whose career at the Anti-Vivisection Hospital had ended so precipitately), the latest of which, *Clinical Medical Discoveries* (1961), described some of the many medical advances that had been made without vivisection. Bayly was the most prominent of the few doctors still working for the anti-vivisection cause in the post-war period, and his writings were lucid and well argued, but his lists of advances that had been made without animal experiments could be no more conclusive that the RDS's lists of advances made with them. How the development of medicine would have been different had vivisection never been permitted is a question of hypothetical history that is unlikely ever to be definitively answered.

The BUAV's latent pacifism resurfaced in the Cold War years, when it renewed its protests against the use of laboratory animals for military research, details of which were not declared in Home Office statistics for reasons of national security. Both the BUAV and the RSPCA noted with concern the use of monkeys and other animals in American rocket tests, and wanted to ensure these were not reproduced in Britain: according to the BUAV, the true objective of the 'conquest of space' was to achieve military supremacy by placing nuclear missiles in orbit.[55] The public, however, were mostly on the side of scientific progress, and watched developments in the 'space race' between the USSR and the West with interest.

In 1957, the Russian dog Laika (Barker) became world famous as the first living creature to orbit the earth. The Soviet government claimed she had been euthanised after 5 days in space, before her oxygen ran out, though she had actually died of overheating within a few hours of launch. Although the exact mode of Laika's death was not known in Britain until secret material was declassified in 2002, this had obviously been a lethal experiment upon a cooperative domestic animal, and it is significant that reports in the British press were overwhelmingly favourable, despite the experiment having been performed by a political rival and nominal enemy on the other side of the 'iron curtain'. The training of 'space dogs' included being confined in ever smaller cages and spun in centrifuges to accustom them to conditions inside a space capsule. It would not have taken much journalistic imagination to make a comparison between the fate of Laika and her fellow space dogs and that of the unwilling human victims of the relentless communist pursuit of technological and industrial superiority over the West. Instead, there was praise, without irony, for the 'selfless contribution' that dogs were making to scientific progress.[56] Indeed, if Laika had been any other species it seems unlikely that any protest would have been made at all, but some feeling that dogs were entitled to special consideration remained: the National Canine Defence League called on all dog owners to observe a minute's silence, and a few protestors gathered outside the Russian embassy, including the 79-year-old Lizzy Lind af Hageby.

The Littlewood Report and After

The RSPCA was still raising concerns over inadequate controls on vivisection in the 1960s: by this time there were over three million procedures annually, six thousand licensed vivisectors, and still only six inspectors, all medical men.[57] In May 1963, the Home Secretary finally responded to pressure and set up a committee, under the chairmanship of lawyer Sir Sydney Littlewood (1895–1967), with members drawn from science, the church, politics and anti-vivisection, to make a thorough review of the law on animal experimentation. Early on in its deliberations, the committee decided that 'vivisection' was no longer the appropriate word for the uses to which most laboratory animals were now being put. While the pro-vivisectionists on the committee may have preferred a more euphemistic term in order to avoid the visceral response that the suggestion of being cut up alive elicits, they were right to point out that the experiences of laboratory animals had changed beyond recognition since the 1876 Cruelty to Animals Act. The solitary physiologist who risked public disapproval to make great discoveries was a world away from the soulless, protocol-driven laboratories that consumed thousands of animals in 'routine' tests.

The committee received little response from the public, and had to rely instead on interviewing the editors of national newspapers to get a sense of public opinion. The editors confirmed that, apart from a few tireless correspondents who wrote into express the same pro- or anti-vivisection sentiments every time the subject was mentioned, they received few letters about it from their readers.[58] The public's desire for improved laboratory animal welfare legislation had long since passed. When the Littlewood report was completed, in 1965, it ran to 255 pages and made 83 recommendations. The report upheld the government's line that there was no overuse of animals and that the licensing system was adequate, though they did recommend that the 1876 Act, and its administration, be overhauled, and made some practical suggestions: the 'debarking' of laboratory dogs by cutting their vocal cords was to cease; Home Office inspectors were to have greater powers and better training; more inspectors were to be recruited; and the public were to be allowed to see animals under experiment.[59] The report was, however, never properly debated, and indeed it was 1971 before parliament discussed it at all. Anti-vivisection groups considered it a whitewash, and Risdon showed that public support could still be mobilised with a little effort by presenting a 300,000-signature petition of protest to parliament, but it was too late to make a difference.[60]

To meet the requirement for experimentation on an ever-larger scale, animals were still being illicitly supplied to laboratories. The RSPCA told the Littlewood committee that there was:

> A thriving trade in procuring and disposing of animals to hospitals and laboratories. But in all too many instances a certain duplicity is practised and members of the public are misled by vaguely-worded advertisements inserted in local newspapers and tending to create the impression that the dealers concerned are genuinely seeking to place unwanted animals with new owners.[61]

Even the RDS admitted that breeders and suppliers were struggling to keep pace with the burgeoning demand: speaking for the Society, Dr Lane-Petter complained of an 'embarrassing lack of animals in this country for trying out all manner of vital new drugs'.[62]

In 1967, the Medicines Control Agency was established, and by 1970, five times more experiments were being performed than in 1946, the great majority for regulatory purposes, and in particular, LD50 testing. Paradoxically, the methodological flaws that made LD50 an imperfect means of predicting human toxicity (differences in reaction between species; limited genetic diversity among rats and mice bred for the laboratory) only served to increase the number of animals used. What was needed, said the regulators, were more tests on an ever-greater variety of species. The thalidomide tragedy in 1959–1961 led to a redoubling of animal testing, which was made mandatory in 1969. Of course, it could never be established for certain that a drug was safe until it was given to patients, but the regulators required such large amounts of 'pre-clinical' animal toxicity data before researchers could even begin to test a drug on humans, that the *British Medical Journal* complained that over-regulation was delaying the introduction of new drugs.[63]

Conclusion

We come to the end of our chronological survey of the anti-vivisection movement at what can only seem an inauspicious period in its history. The total number of animals being used was at an all-time high, and anti-vivisection campaigners lacked the unity and the influence to translate a latent public dislike of vivisection into effective protest, still less to effect a change of heart where it mattered: in government, the medical profession and academia. Mindful of the great deal that remained to be done, one might be tempted to dismiss a century of anti-vivisection activity as having led nowhere. The movement's accomplishments, however, were far from negligible: imperfect and out-dated though the British legislation was, it was still the most comprehensive in the world, and for more than a 100 years the use of animals in scientific research had never been off the ethical and political agenda.

In no other country, over a sustained period, had so much time and effort been devoted to deliberating on the rights and wrongs of animal experimentation. Of course, there were ulterior motives on both sides, from anti-vivisection hospitals hoping to draw funds away from the voluntary sector, to research institutes whose hegemony depended on the supremacy of laboratory experimentation, and both were guilty of manipulating evidence, politicking, and sometimes downright intimidation; there was, however, at the heart of both pro- and anti-vivisection campaigns, a desire to do the right thing, and a firm belief in the importance of their own convictions.

Underlying their sometimes irreconcilable differences was a fundamental disagreement over what science was, or ought to be, since the anti-vivisection movement was founded on the premise that true progress could never come at the expense of cruelty, and so vivisection could not possibly benefit humanity, since it was intrinsically inhumane. It is a difference still to be resolved, and which keeps the opposing parties from achieving anything like mutual understanding. To many of the recipients of a 'scientific' education, anti-vivisection seems a misguided attempt to introduce sentiment and emotion into a sphere where they simply do not belong. To anti-vivisectionists, however, 'nothing which is ethically wrong can ever be scientifically right'.[64] Perhaps, having learned their history, even the most hardened experimentalist will grant them 'some credit for humanity'.[65]

Notes

1. 'Medical notes in parliament', *BMJ*, *1* (1934), 1098.
2. Norah Elam, *The Medical Research Council: What it is and How it Works* (London: LPAVS, 1935) (Elam 1935).
3. Julie V. Gottlieb, *Feminine Fascism: Women in Britain's Fascist Movement, 1923–45* (London: I.B. Tauris, 2003), 299 (Gottlieb 2003).
4. Stuart, *Bloodless Revolution*, 442–444.
5. J.L. Risdon, *Black Shirt and Smoking Beagles: the Biography of Wilfred Risdon, an Unconventional Campaigner* (Scarborough: Wilfred Books, 2013), 340. Mosley replaced Risdon with William Joyce, 'Lord Haw-Haw', in 1934 (Risdon 2013).
6. Gottlieb, *Feminine Fascism*, 64 (2003) (Gottlieb 2003).
7. Richard Thurlow, *Fascism in Britain: a History, 1918–1945* (London: I.B. Tauris, 1998), 148. Kean, *Animal Rights*, 258, states that LPAVS members were 'totally committed' to anti-vivisection (Thurlow 1998).
8. Risdon, *Black Shirt*, 396 (2013) (Risdon 2013).
9. *Anti-Vivisection News-Sheet*, July 1940, 1.
10. 'Victory is ours' (editorial), *Anti-Vivisection News-Sheet*, July 1940, 3; 'We are not amused', *Anti-Vivisection News-Sheet*, July 1940, 19 (Victory is Ours 1940; We are not Amused 1940).

11. Guy Coleridge, 'Dog murder in Germany', *Anti-Vivisection News-Sheet*, July 1940, 19 (Coleridge 1940).

12. Nadja Durbach, *Bodily Matters: The Anti-Vaccination Movement in England, 1853–1907* (Durham: Duke University Press, 2005), 232 (Durbach 2005).

13. 'Vivisection—a depressed industry', *Animals' Defender*, 60 (1940), 1; Hopley, *Campaigning Against Cruelty*, 11 (Vivisection—A Depressed Industry 1940).

14. 'Very "hush-hush"', *Anti-Vivisection News-Sheet*, September 1940, 1; 'Our law allows this', *Animals' Defender*, 61 (1941), 1 (Very "Hush-Hush" 1940; Our Law Allows This 1941).

15. 'Blast!' (editorial), *Antivivisection News-Sheet*, September 1940, 2 (Blast 1940).

16. 'Estimate the suffering', *Animals' Defender*, 60 (1941), 1 (Estimate the suffering 1941).

17. M. Beddow Bayly, 'Recent experiments in British Laboratories', *Anti-Vivisection News-Sheet*, October 1953, 72 (Bayly 1953).

18. '"Interesting" experiments', *Animals' Defender*, 60 (1941), 1; Hopley, *Campaigning Against Cruelty*, 51 ("Interesting" Experiments 1941).

19. They also campaigned for a ban on the kosher method of killing, a not uncommon demand at the time: E.J.E., 'The History of Torture Through the Ages' (review), *Antivivisection News-Sheet*, October 1940, 2 (E.J.E 1940).

20. Letter, H. Marrian Perry to Sir Leonard, 10 July 1940, Well SA/RDS/C2.

21. Memo, 10 July 1940, Well SA/RDS/C2.

22. Memo from the honorary secretary of the RDS, September 1942, Well SA/RDS/C2.

23. RDS memoranda: 7 June 1943, 217; 20 January 1941, 207, Well SA/RDS/C2.

24. RDS archives, November 1945, 259, Well SA/RDS/C2.

25. RDS archives, 26 April 1946, 233, Well SA/RDS/C2.

26. *Times Law Reports*, *11* (1895), 541.

27. Peter Radan, 'Antivivisection and Charity', *Sydney Law Review*, *35* (2013), 519–539 (Radan 2013).

28. Bristol physiology department order books, Well GC/108/1; Risdon, *Black Shirt*, 346, 369.

29. Hopley, *Campaigning Against Cruelty*, 54–55.

30. Vyvyan, *The Dark Face*, 159.

31. Boria Sax, *Animals in the Third Reich: Pets, Scapegoats, and the Holocaust* (New York: Continuum, 2000) (Sax 2000).

32. A Universal Declaration of Animal Welfare was drafted in 2000 but has yet to be endorsed by the United Nations.

33. 'Deputation to the Home Secretary', *Animals' Defender*, April 1948, 47; Ryder, *Victims of Science*, 224 (Deputation to the Home Secretary 1948).

34. Ryder, *Victims of Science*, 226.

35. 'Do they ever say "no"?', *Animals' Defender*, March 1954, 46. I have been unable to discover evidence of a licence being refused under the 1876 Act (Do They Ever Say "No" 1954).

36. 'Atomic weapon experiments', *Animals' Defender*, 68 (1948), 1. The following year, the first historical account of the subject was published, E. Westacott's *A Century of Vivisection and Anti-Vivisection* (Ashingdon: C.W. Daniel, 1949) (Westacott 1949; Atomic Weapon Experiments 1948).

37. 'What happens at Harwell', *Animals' Defender*, January 1954, 1 (What Happens at Harwell 1954).

38. A.J. Eagleton, 'The standardization of tuberculin', *Lancet*, 1 (1921), 429–431 (Eagleton 1921).

39. J.H. Gaddum, 'John William Trevan, 1887–1956', *Biographical Memoirs of Fellows of the Royal Society*, 3 (1957), 273–288 (Gaddum 1957).

40. J.W. Trevan, 'The error of determination of toxicity', *Proceedings of the Royal Society of London*, 101 (1927), 483–514 (Trevan 1927).

41. 'The prohibition of vivisection', *Lancet*, 2 (1930), 1360.

42. Harry Eckstein, *Pressure Group Politics: The Case of the British Medical Association* (Stanford, CA: Stanford University Press, 1960), 75–78 (Eckstein 1960).

43. Marian Bielschowsky and H.N. Green, 'Fractionation, chemical properties and effective doses', *Lancet*, 2 (1943), 1531–1535 (Bielschowsky and Green 1943).

44. G.R. Cameron and R.F. Milton, 'Toxicity of tannic acid: an experimental investigation', *Lancet*, 2 (1943), 179–186 (Cameron and Milton 1943).

45. 'Antidote to arsenicals', *Lancet*, 2 (1945), 854–855.

46. 'Correspondence With a vivisector', *Animals' Defender*, June 1952, 7 (Correspondence With a Vivisector 1952).

47. Obituary, *BMJ* (2004), http://www.bmj.com/content/suppl/2004/04/15/328.7445.960-e.DC1, viewed 24 June 2016. Dwindling membership led the LPAVS to merge with the NAVS in 1957. Risdon became secretary of the merged organization, and worked towards unity with the BUAV up to his death in 1967, though as the BUAV remained

committed to total abolition as the only option, unity was never achieved: Risdon, *Black Shirt*, 496 (Obituary 2004).

48. 'Books reviewed', *Antivivisection News*, January 1953, 2 (Books Reviewed 1953).
49. Hopley, *Campaigning Against Cruelty*, 57–59.
50. Risdon, *Black Shirt*, 493, 659–668 (2013) (Risdon 2013).
51. Ryder, *Victims of Science*, 224–225.
52. Ryder, *Victims of Science*, 226.
53. C.W. Hume, 'Soldiers and laboratory animals: an analogy for experimental biologists', *Lancet*, *271* (1958), 424–426.
54. W.M.S. Russell and R.L. Burch, *The Principles of Humane Experimental Technique* (London, Methuen, 1959).
55. 'Those rocket tests', *Animals' Defender*, *72* (1952), 1 (Those Rocket Tests 1952).
56. Tacium, 'A history of anti-vivisection… Part I'.
57. Niven, *History of the Humane Movement*, 86.
58. John Bleby, 'The Littlewood Committee report on experiments on animals', *Journal of Small Animal Practice*, 7 (1966), 205–214 (Bleby 1966).
59. Ibid.
60. Risdon, *Black Shirt*, 498 (2013) (Risdon 2013).
61. *Report of the Departmental Committee on Experiments on Animals* (HMSO, 1965), 177 (HMSO 1965).
62. *The Anti-Vivisectionist*, July/August 1965, 44.
63. Bleby, 'The Littlewood Committee'.
64. 'Animals and us: quotations', *Quest*, 89 (2001), https://www.theosophical.org/publications/quest-magazine/42-publications/quest-magazine/1325-animals-and-us-quotations, viewed 24 June 2016 (Animals and Us: Quotations 2009).
65. Robert Knox, 'Xavier Bichat: his life and labours; a biographical and philosophical study', *Lancet*, 2 (1854), 393–396 (Knox 1854).

References

Animals and Us: Quotations. (2009). *Quest,* 89. Retrieved June 24, 2016, from https://www.theosophical.org/publications/quest-magazine/42publications/questmagazine/1325-animals-and-us-quotations.

Atomic Weapon Experiments. (1948). *Animals' Defender, 68,* 1.

Bayly, M. B. (1953, October). Recent experiments in British Laboratories. *Anti-Vivisection News-Sheet*, 72.

Bielschowsky, M, & Green, H. N. (1943). Fractionation, chemical properties and effective doses. *Lancet, 2*, 1531–1535.

Blast! (Editorial). (1940, September). *Antivivisection News-Sheet*, 2.

Bleby, J. (1966). The Littlewood Committee report on experiments on animals. *Journal of Small Animal Practice, 7*, 205–214.

Books Reviewed. (1953, January). *Antivivisection News*, 2.

Cameron, G. R., & Milton, R. F. (1943). Toxicity of Tannic acid: An experimental investigation. *Lancet, 2*, 179–186.

Coleridge, G. (1940, July 19). Dog murder in Germany. *Anti-Vivisection News-Sheet*.

Correspondence With a Wivisector. (1952, June). *Animals' Defender, 7*.

Deputation to the Home Secretary. (1948, April). *Animals' Defender, 47*.

Do They Ever Say "No"?. (1954, March). *Animals' Defender, 46*.

Durbach, N. (2005). *Bodily matters: The anti-vaccination movement in England, 1853–1907*. Durham: Duke University Press.

Eagleton, A. J. (1921). The standardization of tuberculin. *Lancet, 1*, 429–431.

E.J.E. (1940, October). The history of torture through the ages (review). *Antivivisection News-Sheet*, 2.

Eckstein, H. (1960). *Pressure group politics: The case of the British Medical Association*. Stanford, CA: Stanford University Press.

Elam, N. (1935). *The medical research council: What it is and how it works*. London: LPAVS.

Estimate the Suffering. (1941). *Animals' Defender, 60*, 1.

Gaddum, J. H. (1957). John William Trevan, 1887–1956. *Biographical Memoirs of Fellows of the Royal Society, 3*, 273–288.

Gottlieb, J. V. (2003). *Feminine fascism: Women in Britain's fascist movement, 1923–1945* (p. 299). London: I.B. Tauris.

HMSO. (1965). *Report of the Departmental Committee on Experiments on Animals*, 177.

Hume, C. W. (1958). Soldiers and laboratory animals: An analogy for experimental biologists. *Lancet, 271*, 424–426.

"Interesting" Experiments. (1941). *Animals' Defender, 60*, 1.

Knox, R. (1854). Xavier Bichat: His life and labours: A biographical and philosophical study. *Lancet, 2*, 393–396.

Obituary. (2004). *BMJ*. Retrieved June 24, 2016, from http://www.bmj.com/content/suppl/2004/04/15/328.7445.960e.DC1.

Our Law Allows This. (1941). *Animals' Defender, 61,* 1.

Radan, P. (2013). Antivivisection and charity. *Sydney Law Review, 35,* 519–539.

Risdon, J. L. (2013). *Black shirt and smoking beagles: The biography of Wilfred Risdon, an unconventional campaigner.* Scarborough: Wilfred Books.

Russell, W. M. S., & Burch, R. L. (1959). *The principles of humane experimental technique.* London: Methuen.

Sax, B. (2000). *Animals in the Third Reich: Pets, scapegoats, and the holocaust.* New York: Continuum.

Those Rocket Tests. (1952). *Animals' Defender, 72,* 1.

Thurlow, R. (1998). *Fascism in Britain: A history, 1918–1945* (p. 148). London: I.B. Tauris.

Trevan, J. W. (1927). The error of determination of toxicity. *Proceedings of the Royal Society of London, 101,* 483–514.

Very "Hush-Hush". (1940, September). *Anti-Vivisection News-Sheet,* 1.

Victory is Ours (Editorial). (1940, July). *Anti-Vivisection News-Sheet,* 3.

Vivisection—A Depressed Industry. (1940). *Animals' Defender, 60,* 1.

We are not Amused. (1940, July). *Anti-Vivisection News-Sheet,* 19.

Westacott's, E. (1949). *A century of vivisection and anti-vivisection.* Ashingdon: C.W. Daniel.

What Happens at Harwell. (1954, January). *Animals' Defender,* 1.

8

Conclusion

I hope that what I have said respecting the exercise of humanity to animals, will awaken your attention to that virtue. Neither punishment, indeed, nor reward, are any where held out as inducements to its practice; but it is therefore not less a virtue, and you will have the satisfaction, at any rate, of doing good for its own sake, a thing, I fear, of not common occurrence in the present constitution of things.
James Lawson Drummond, Letters to a Young Naturalist (1831)

It might be thought that the last thing that twenty-first century Britain needs is more feeling for animals. The popular press is awash with sentimental animal stories of love and devotion and/or cruelty and neglect, designed to tug at our heartstrings or arouse angry indignation. Of course, we hardly expect scientists to be subject to such passions. Science is a privileged enclave from which normal emotional responses are excluded, and has been since at least the early-twentieth century, when it became acceptable for its practitioners to claim that, as Shaw stated ironically, 'as a Man of Science you are beyond good and evil'.[1] Thus, while no scientist wanted to perform vivisection, those who considered it necessary refused to allow personal feelings to get in the way. The ability to suppress one's natural sense of compassion became

© The Author(s) 2017
A.W.H. Bates, *Anti-Vivisection and the Profession of Medicine in Britain*,
The Palgrave Macmillan Animal Ethics Series, DOI 10.1057/978-1-137-55697-4_8

something of a test for those embarking on a career in medical science, and they seem at times to have been proud of the objectivity they managed to achieve; or perhaps, as Vyvyan put it, 'scientists will go to any length to avoid feeling what they know'.[2]

Since the nineteenth century, the main justification for vivisection has been utility. The traditional Christian teachings that animals lack souls, and that human dominion over them is divinely ordained, certainly did not help their cause, but any tendency for these doctrines to encourage the exploitation of laboratory animals has to be weighed against the efforts of many dedicated anti-vivisection campaigners who were inspired by their Christian faith. The laboratory animal's tormentors were materialistic utilitarians, not Bible-believing Christians.

My judgement on their experiments, for what it is worth, is that the vast majority would have failed present day utilitarian criteria such as those proposed by Singer.[3] Of course, some were medically useful, a few led to very significant developments in medical practice, and only rarely (despite what anti-vivisectionists liked to claim) did extrapolation of results from animals to humans lead to dangerous error—for the most part, laboratory mammals tend to behave physiologically in a way very similar to ourselves. Most experiments on animals carried out during the period covered by this book were not, of course, groundbreaking scientific studies, but demonstrations, repeat experiments, and routine tests. With regard to novel investigative research on animals, however, the historical record does nothing to substantiate extreme views that either it reliably produced medical breakthroughs, or that no good ever came of it.

What the history of the anti-vivisection movement does demonstrate, is that utility has not always been the main issue, and that many early anti-vivisectionists ignored it altogether. Their motivation lay in their concerns that vivisection would exert a demoralising effect on individual experimenters, and on society as a whole. The potential benefits to medical knowledge, and whether the animals used were physiologically, intellectually, or spiritually comparable to ourselves, were issues of lesser importance than the feeling that inflicting pain on helpless creatures was a morally dangerous business. It was the fear that vivisection would promote a more brutal society that united people of diverse backgrounds in opposition to it, from the poor who feared being experimented on themselves, to the rich, who worried that the example of cruelty set by the professional classes might spread to the unlearned and precipitate them down a slippery slope to moral anarchy.

At a time when the opponents of vivisection concern themselves almost exclusively with the rights and interests of animals (a subject with its own lengthy history), it is salutary to recall that the radical Animals' Friend Society's five objections to vivisection did not mention animals at all: according to them, it was a moral failing, created public animosity against scientists, fostered cruelty towards humans, diverted charity away from human causes, and offended God. On these principals was built an anti-cruelty movement unequalled anywhere in the world.

Of the five objections, two might currently be accepted without demur: vivisection has certainly created animosity against scientists (some of whom have been the victims of violent attacks) and it probably diverts charitable efforts away from human causes (for some reason, it is anti-vivisectionists who tend to be blamed for this). The claim that vivisection promotes cruelty to humans is perhaps best regarded as unproven, though the link between cruelty to animals in general and violence towards humans is now well established. There has been no official pronouncement from any major religious denomination on whether vivisection offends God, despite the efforts, in recent years, of a growing number of animal theologians. What is curious is that the objection that vivisection is a moral failing on the part of the perpetrator, which was first on the list in the nineteenth century, is, nowadays, probably the most likely to be overlooked.

How did something that was once so important come to be so neglected? The declining interest in virtue ethics, especially in academic circles (where it is now enjoying something of a renaissance), has perhaps been partly responsible, as has an increased focus on the animals themselves, and in particular their rights and interests. So dominant has this approach become that it now seems somewhat strange to worry about the effect that vivisection has on us as humans.

One advantage of taking an anthropocentric view is that it does not matter whether animals have rights, or even feelings. Modern environmental ethics has, after all, been built on a foundation of human virtue: plants and landscapes do not have rights or feelings, but it is wrong to destroy them selfishly to serve one's own interests. When we hear of the environment being thoughtlessly damaged, we may well ask ourselves 'what sort of person would do a thing like that?'[4] It is the question that nineteenth-century anti-cruelty campaigners demanded of vivisectionists. At the very least, they expected them to undertake their work in a spirit of honesty, humility and mercy, and not be casual, uncaring, self-righteous or cruel.[5]

Apart from very rare cases of experimenters, notably Magendie, who did seem to be thoroughly heartless, the vast majority of scientists were, and are, determined to behave responsibly. They did not enjoy, and may positively have disliked, vivisection, and could suppress their natural emotional response to it only because a dispassionate attitude that would have been considered callous in everyday life was permitted, and even expected, among medical scientists. For a profession that allowed its practitioners to set conventional moral norms aside, Vyvyan's question—'To whom or what is a scientist responsible?'—became a crucial one.[6]

The practical answer was 'to him- or herself', because medical scientists, as educated people with privileged access to knowledge, were expected to self-regulate, an oxymoronic concept that led to the carte blanche conclusion that nothing could be cruel provided it was scientific. For many anti-vivisectionists, the exact reverse was true: morality came first, and nothing morally wrong could be scientifically right. For them, the vivisectors' claim that they were excused from moral guilt because they were engaged in science was an inversion of the proper order: the deep understanding of the world that constituted 'true' science would, they believed, preclude its possessors from trying to wrest nature's secrets out of her by brute force.[7]

We may now take it for granted that scientists are detached, objective, and unemotional, and that they are permitted to do things in the laboratory that would be socially unacceptable, illegal, or even damnable if done outside it, but in the nineteenth and early-twentieth centuries, when the philosophical and ethical rules of conduct for laboratory medicine were being laid down, it was not a foregone conclusion that amoral (or immoral) materialism would win out. In fact, the acceptance of animal experimentation in medicine represented an ethical break with the past, as the compassion and sensitivity that had characterised the medical gentleman were supplanted by the persona of the cool, impassive, white-coated medical scientist. Indeed, in a remarkable volte-face, emotion came to be seen as self-indulgent and unmanly, feelings as undesirable, and intuition as a methodological failing. Organizations such as the OGA that sought to preserve the role of the spirit and emotions in science became an increasingly marginalised voice, part of an alternative (sub)culture of utopian reformers, vegetarians, visionaries, and radicals.

The obvious heirs to this countercultural movement were the hippies of the 1960s and 1970s, with their ethos of pacifism, spirituality, syncretism and environmentalism, and there is enough common ground—the influence of transcendentalism, Eastern religions, Swedenborgianism, Unitarianism and theosophy, and the granting of equivalent legitimacy to laboratory experiments and subjective experience—to see the new age movement of the 1960s as a resurgence of the programme pursued by Oldfield and other reformers almost a century before, revived by a post-war generation after memories of the unfortunate association between the back-to-nature movement and National Socialism had faded. Anti-vivisection was revived too, though its wartime taint of disloyalty and subversion had perhaps not entirely been forgotten.

From the early history of British anti-vivisection emerge two points of relevance to us today: that medical practitioners thought from the first that vivisection was incompatible with the humane ethos of their profession, and that the objectivity of science was a contested construct. Medical practitioners in the nineteenth century *felt* that vivisection was wrong because it repelled them. They were prepared to accept emotions as evidence, and to situate their professional work in the wider context of their beliefs, something that the promoters of a morally neutral laboratory culture were determined to make unacceptable.

Theirs, of course, was the view that prevailed: there never were many medics who actually performed vivisection, but all were taught in medical schools that it was indispensable for knowledge, and that those who opposed it were enemies of science. To speak out was disloyalty, and medical students and young researchers (as I know from experience) went along with the culture of animal experimentation because to dissent was heresy. It may encourage future dissenters to note that the conception of science as beyond morality is a comparatively recent, debatable, and perhaps ephemeral, one. Possibly no one ever can truly reach a state where they could look upon vivisection unfeelingly, and if they could, they would have lost their humanity.

For ethicists, the most important lesson from history is that it *is* possible to construct a coherent and effective case against vivisection in which neither utilitarianism nor animal rights needs feature prominently!

Perhaps, in my lifetime, vivisection will be confined to history. Attitudes constantly change, and within living memory, experiments were carried out in the name of science (perhaps science is not such a civilising influence after all) that most scientists would find unacceptable, even abhorrent, today. Moral judgements cannot be reliably applied retrospectively; since the nineteenth century, the scope and influence of medical science has increased beyond all bounds, and there will always be fresh ethical challenges to be faced. It takes time to find the right answers, particularly when our capacity to do surpasses our capacity to know. The vigour with which our predecessors engaged in the vivisection debate was a sign of an impressive intellectual and moral commitment from participants on both sides to do the right thing. We owe them our gratitude, and their arguments our attention.

Notes

1. George Bernard Shaw, 'These scoundrels!,' *Sunday Express*, 7 August 1927 (Shaw 1927).
2. Vyvyan, *In Pity*, 20.
3. Singer, 'Animals and the Value of Life'.
4. Matt Zwolinski and David Schmidtz, 'Environmental virtue ethics: what it is, and what it needs to be', in Daniel C. Russell (ed.) *The Cambridge Companion to Virtue Ethics* (Cambridge: Cambridge University Press, 2012), 221–239, on p. 224 (Zwolinski and Schmidtz 2012).
5. Compare Oakley's discussion of induced abortion: 'Virtue ethics', 209.

6. Vyvyan, *In Pity*, 44.
7. Li, 'An unnatural alliance?'.

References

Drummond, J. L. (1831). On humanity to animals. *Edinburgh Philosophical Journal, 13*, 172–183.

Shaw, G. B. (1927, August 7). These scoundrels! *Sunday Express.*

Zwolinski, M., & Schmidtz D. (2012). Environmental virtue ethics: What it is, and what it needs to be. In D. C. Russell (Ed.), *The Cambridge companion to virtue ethics* (pp. 221–239, on p. 224). Cambridge: Cambridge University Press.

Chronology of Events

1792 Thomas Taylor publishes *A Vindication of the Rights of Brutes*
1809 Lord Erskine introduces an unsuccessful Cruelty to Animals Bill
1822 Richard Martin's Cruel and Improper Treatment of Cattle Act passed
1824 François Magendie gives vivisection demonstration in London; SPCA founded
1829 Anatomy murders in Edinburgh
1831 Marshall Hall publishes ethical guidelines for vivisectionists
1832 Anatomy Act
1833 Lewis Gompertz resigns from SPCA and founds Animals Friend Society
1835 Anti-cruelty law extended to domestic animals
1842 Etherington publishes *Vivisection Investigated and Vindicated*
1847 Marshall Hall publishes guidelines for vivisectors in the *Lancet*; Vegetarian Society founded
1852 Gompertz publishes *Fragments in Defence of Animals*
1858 Medical Act introduces medical registration
1859 Charles Darwin publishes *On the Origin of Species*

© The Editor(s) (if applicable) and The Author(s) 2017   **203**
A.W.H. Bates, *Anti-Vivisection and the Profession of Medicine in Britain*,
The Palgrave Macmillan Animal Ethics Series, DOI 10.1057/978-1-137-55697-4

1860   Mary Tealby opens the Temporary Home for Lost and Starving Dogs in Holloway

1873   John Burdon-Sanderson publishes *Handbook for the Physiological Laboratory*

1874   Physiologist Éugène Magnan gives demonstration in Britain

1875   Attempted prosecution of Magnan for vivisection; first Royal Commission on Vivisection; Victoria Street Society (later the National Anti-Vivisection Society) founded by Frances Power Cobbe, who publishes *The Moral Aspects of Vivisection*; Archbishops of York and Dublin sign memorial against vivisection

1876   Cruelty to Animals Act ('Vivisection Act') passed; Physiological Society founded; London Anti-Vivisection Society Founded

1880   Anna Kingsford receives Paris MD, obtained without vivisection

1881   The International Medical Congress passes a resolution that vivisection by 'competent persons' should not be restricted

1882   Order of the Golden Age founded

1885   John Ruskin resigns his Oxford professorship in protest at vivisection

1891   National Canine Defence League founded; Henry Salt founds the Humanitarian League

1898   Frances Cobbe leaves the NAVS after it compromises on abolition, and founds the British Union for the Abolition of Vivisection; Oldfield opens first anti-vivisection hospital

1901   Stephen Coleridge publishes *The Metropolitan Hospitals and Vivisection*

1902   (Imperial) Cancer Research Fund Founded

1903   'Brown Dog' statue placed in Battersea; Lizzy Lind af Hageby and the Duchess of Hamilton found the Animal Defence and Anti-Vivisection Society; Lind af Hageby publishes *The Shambles of Science*; Professor Bayliss of UCL successfully sues Coleridge for libel; Battersea (National) Anti-Vivisection Hospital opens

1904   St Francis's Anti-Vivisection Hospital closes; OGA reconstituted

1905   Dogs Protection Bill first introduced in parliament

1906   Second Royal Commission on Vivisection begins; Dogs Act protects strays from vivisection; Dogs Protection Bill to ban all experiments on dogs is supported by a 300,000-signature petition but defeated in parliament
1907   Brown Dog riots begin
1908   Research Defence Society founded
1909   Wistar rat developed especially for experimental use
1910   Brown Dog statue removed
1911   National Insurance Act
1912   Second Royal Commission reports
1913   Medical Research Committee and Advisory Council founded; Dogs Protection Bill defeated in committee
1915   War Department Experimental Ground set up at Porton Down
1917   Association for the Advancement of Medical Research amalgamates with the RDS
1920   Medical Research Council receives royal charter; Humanitarian League closes down
1923   British Empire Cancer Campaign founded
1924   First Labour government
1926   University College professor acquitted of dog stealing; Major Charles Hume founds the University of London Animal Welfare Society
1927   BMA conference on research and animal experimentation suggests using strays for vivisection; LD50 testing developed
1928   Grove-Grady case: money given to help animals denied charitable status
1930   Joseph Kenworthy introduces a bill to ban spending public money on vivisection
1933   German government bans vivisection
1935   National Anti-Vivisection Hospital closes; Norah Elam publishes *The Medical Research Council, What It Is and How It Works*
1938   ULAWS becomes the Universities Federation for Animal Welfare (UFAW); OGA ceases activity in Britain
1939   Last official figures list over 900,000 animal experiments a year
1940   Norah Elam and Wilfred Risdon imprisoned
1941   Anti-Vivisection ruled not charitable

1942   Wilfred Risdon becomes secretary of LPAVS

1943   First research using LD50 published in Britain

1944   Draize test devised in USA

1946   National Health Service Act

1947   NAVS loses appeal to have its charitable status restored

1948   Universal Declaration of Human Rights

1950   Diseases of Animals Act

1952   First prosecution for liberation of laboratory animals

1954   World congress of anti-vivisection societies meets in London

1956   Therapeutic Substances Act passed

1957   The London and Provincial Anti-Vivisection Society amalgamates with the NAVS; Laika dies during space mission; NCDL campaigns against the use of dogs in space research

1959   William Russell and Rex Burch publish *The Principles of Humane Experimental Technique* and propose the three R's

1965   Littlewood Committee reports

1967   Medicines Control Agency established

1969   Testing of drugs on animals becomes mandatory

1971   Oxford philosophers publish *Animals, Men and Morals*

Bibliography

Bamwell, A. (1994). *Ecology in the 20th century: A history*. New Haven: Yale University Press.

Boddice, R. (2007). *A history of attitudes and behaviours toward animals in eighteenth- and nineteenth-century Britain: Anthropocentrism and the emergence of animals*. Lampeter: Mellen.

Boddice, R. (2011). Vivisecting major: A victorian gentleman scientist defends animal experimentation, 1876–1885. *Isis, 102,* 215–237.

DeMello, M. (2012). *Animals and society: An introduction to human-animal studies*. New York: Colombia University Press.

Feller, D. A. (2009). Dog fight: Darwin as animal advocate in the antivivisection controversy of 1875. *Studies in History and Philosophy of Biological and Biomedical Sciences, 40,* 265–271.

Giubilini, A., & Minerva, F. (2013). After-birth abortion: Why should the baby live? *Journal of Medical Ethics, 39,* 261–263.

Harrison, B. (1982). *Peaceable kingdom: Stability and change in modern Britain*. Oxford: Clarendon Press.

Hursthouse, R. (2001). *On virtue ethics*. Oxford: Oxford University Press.

Kete, K. (1994). *The beast in the boudoir: Pet keeping in nineteenth-century Paris*. Berkeley: University of California Press.

Klein, E., Burdon-Sanderson, J., Forster, M., & Brunton, T. L. (1873). *Handbook for the physiological laboratory*. London: J. and A. Churchill.

© The Editor(s) (if applicable) and The Author(s) 2017
A.W.H. Bates, *Anti-Vivisection and the Profession of Medicine in Britain*,
The Palgrave Macmillan Animal Ethics Series, DOI 10.1057/978-1-137-55697-4

Li, C. (2012). An unnatural alliance? Political radicalism and the animal defence movement in late Victorian and Edwardian Britain. *EurAmerica, 42,* 1–43.

Linzey, A. (1999). *Animal gospel.* Louisville, KY: Westminster John Knox Press.

Merriam, G. (2008). *Virtue ethics and the moral significance of animals.* Ph.D. thesis, Rice University, TX.

Newton, D. E. (2003). *The animal experimentation debate: A reference handbook.* Santa Barbara, CA: ABC Clio.

Phillips, M. T., & Sechzer, J. A. (1989). *Animal research and ethical conflict: An analysis of the scientific literature: 1966–1986.* New York: Springer.

Richardson, R. (2001). *Death, dissection, and the destitute.* London: Phoenix Press.

Rupke, N. A. (Ed.). (1990). *Vivisection in historical perspective.* London: Routledge.

Russell, D. C. (Ed.). (2013). *The Cambridge companion to virtue ethics.* Cambridge: Cambridge University Press.

Sorenson, J. (Ed.). (2014). *Critical animal studies: Thinking the unthinkable.* Toronto: Canadian Scholars' Press.

Stepan, N. (1984). *The idea of race in science: Great Britain 1800–1960.* London: MacMillan.

Turner, J. (1980). *Reckoning with the beast: Animals, pain, and humanity in the Victorian mind.* Baltimore, MD: Johns Hopkins University Press.

Walters, K. S., & Portmess, L. (Eds.). (2001). *Religious vegetarianism: From Hesiod to the Dalai Lama.* New York: State University of New York Press.

Wynter, A. (1869). *Subtle brains and lissom fingers: Being some of the chisel-marks of our industrial and scientific progress, and other papers.* London: Robert Hardwicke.

Index

© The Editor(s) (if applicable) and The Author(s) 2017
A.W.H. Bates, *Anti-Vivisection and the Profession of Medicine in Britain*,
The Palgrave Macmillan Animal Ethics Series, DOI 10.1057/978-1-137-55697-4